TOWPATHS to TUGBOATS

A History of American Canal Engineering

FOREWORD

Much has been written about the Historic Canal Era in the United States and Canada but little about the coordinated efforts of those dedicated men of history, the early Engineers, who made it all happen. In an effort to tell this story, the various Directors of the American Canal Society listed below, have pooled their knowledge and reference material to put together the following account of the canals of antiquity, the great canal-builders of England, the early visionaries in America, the giants of the American Canal Engineering period in the 1800's, and the carry-over of American canal expertise into the Twentieth Century.

Most of the early American canal builders were practical Engineers with little formal training, other than surveying. Many of them were trained on the job, in the backwoods and swamps of New York State, Ohio and Pennsylvania. Physical hardship and sickness discouraged all but the most hardy. In spite of this, the American Canal Engineers wrought marvelous works — planned, organized and supervised the cutting of water communications through the wilderness, and were in general the most highly paid and most respected "professionals" of the Canal Era.

Our accounts of individual activities of the most famous early canal engineers, such as Benjamin Wright, Canvass White, Loammi Baldwin, Nathan Roberts, Edward Gill, Colonel By, William Hamilton Merritt, etc., are limited in length due to the great number of personages we wish to cover. For this, we apologize to the exponents of particular "favorites" and refer them (for further reading and study) to the extensive Bibliography published at the rear of this book.

William H. Shank, P.E.
Author & Publisher
September 1985

Contributing Authors:

Captain Thomas F. Hahn, Ed. D.
T. Gibson Hobbs, Jr.
Robert S. Mayo, P.E.

First Printing - May 1982
Second Printing - Sept. 1985

Published by
The American Canal and Transportation Center
809 Rathton Road, York, Pa. 17403

All rights reserved including the right to reproduce this book or portions thereof in any form, except for the inclusion of brief quotations in a review.

ISBN 0-933788-40-1   Copyright William H. Shank 1982

TABLE OF CONTENTS

Foreword...1
The Canals of Antiquity..3
    The Middle East...3
    China...4
    Greece..4
    Roman Empire..4
    Early Continental European Canals.............................5
    Early British Canals..6
British Canal Builders...7
    James Brindley 1716-1771......................................7
    Thomas Telford 1757-1834......................................8
The Visionaries...11
    Robert Fulton 1765-1815......................................11
    Elkanah Watson 1758-1842.....................................12
    George Washington 1732-1799..................................13
Early Canal Builders in America.....................................15
    John Christian Senf 1753-1806................................15
    Loammi Baldwin 1745-1807.....................................16
    William Weston 1753-1833.....................................16
The Erie Canal..17
    Benjamin Wright 1770-1842....................................19
    James Geddes 1763-1838.......................................20
    Nathan S. Roberts 1776-1852..................................20
    Canvass White 1790-1834......................................22
"Canal Fever"...23
    Massachusetts..23
    Ohio...23
    Pennsylvania...24
    Maryland...26
    New Jersey...27
    Virginia...29
    Indiana..30
    David Stanhope Bates 1777-1839...............................31
    John Bloomfield Jervis 1795-1885.............................32
    Horatio Allen 1802-1889......................................33
    William Milnor Roberts 1802-1882.............................33
    Edward Hall Gill 1806-1868...................................34
    Charles Ellet, Jr. 1810-1862.................................35
    Principal Historic Canals and Their Engineers................36
Canal Engineering...37
    Hydraulic Canal Cement.......................................39
Coal Carrying Canals..40
The Canadian Canals...41
    William Hamilton Merritt 1793-1862...........................42
Illinois and Michigan Canal...44
    William Gooding 1803-1878....................................44
Wisconsin's Portage Canal...48
Army Corps of Engineers...49
St. Mary's Canal..50
    Orlando M. Poe...50
Panama Canal..51
    The French Canal...52
    Theodore Roosevelt 1858-1919.................................53
    William Crawford Gorgas 1854-1920............................55
    The First Chief Engineers....................................55
    George Washington Goethals 1858-1928.........................56
    Construction...57
Navigable Rivers..58
    The Ohio...58
    The Mississippi..60
    The Tennessee..61
The Saint Lawrence Seaway...62
The Tenn-Tom Waterway...65
Historic Canal Preservation...66
Future Inland Water Travel..69
Bibliography..69
Index...71
Publications of the American Canal & Transportation Center..........72

THE CANALS OF ANTIQUITY

By Thomas F. Hahn, Ed. D.

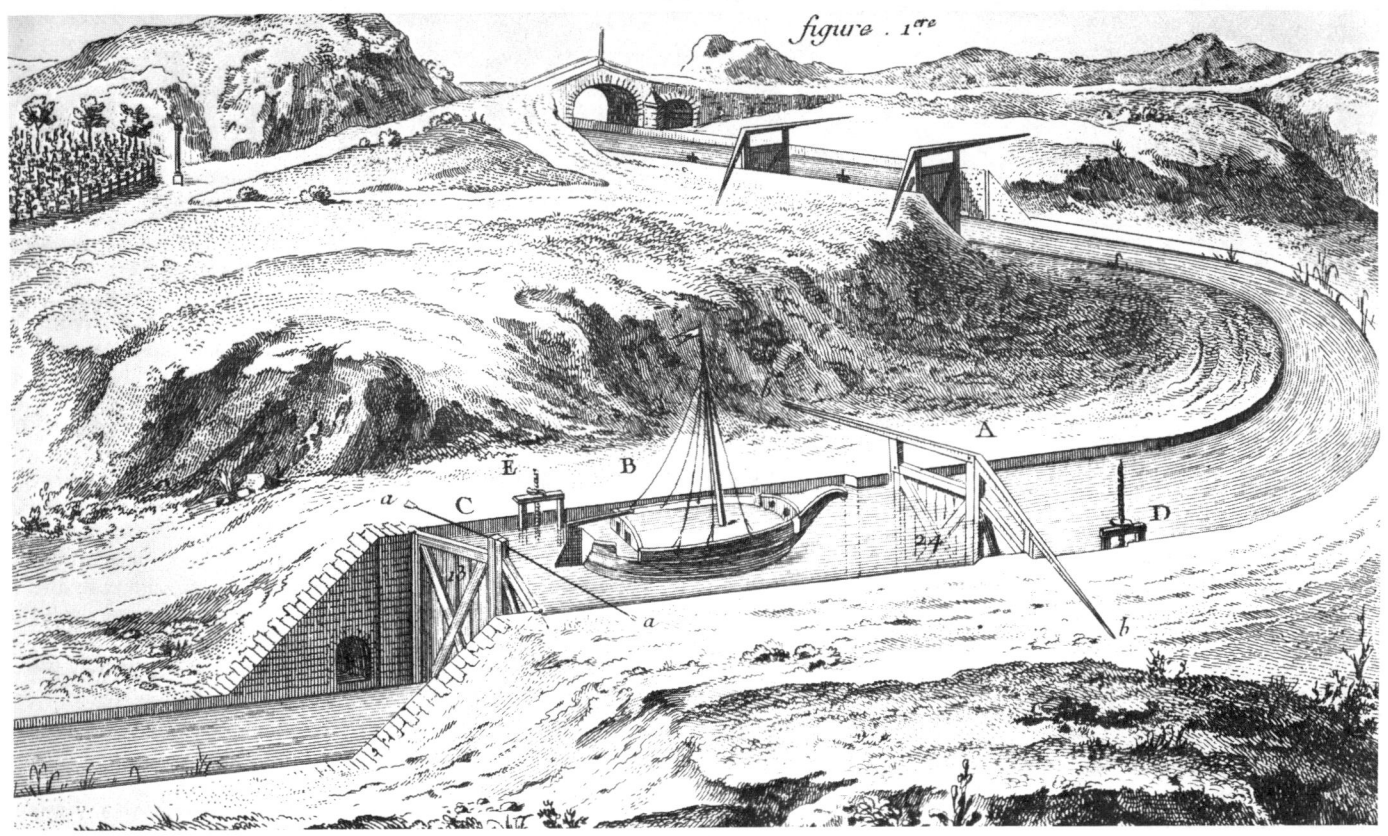

An old print of the Languedoc Canal, or "Canal du Midi," built in South France in 1681 — one hundred fifty miles long, with one hundred locks to connect the Mediterranean with the Atlantic. This illustration, which shows mitre gate locks of the type designed by Leonardo da Vinci, was first published in "Diderot Pictorial Encyclopedia" in 1762. (Courtesy Bob Mayo.)

From the dawn of history, the artificial diversion of water from the rivers, lakes and seas has been closely associated with the rise and fall of different cultures and civilizations. While the early canals were undoubtedly used for irrigation purposes, the old Kings, Pharaohs and Caesars saw the possibilities of artificial waterways for extension of their influence over larger territories and the people in them. Military as well as economic control of entire regions accrued to the benefit of those who controlled the waters. Archeologists have found frequent evidence of ancient, dried-up canals in the Middle East and Europe which were dug, no doubt by slave labor, and abandoned and forgotten with the demise of the Empires which created them.

THE MIDDLE EAST

The first canal builder for whom there is historical documentation (a drawing on an earthen pot) now in the Ashmolean Museum at Oxford, England) was "King Scorpion" of Egypt who built a canal in Upper Egypt ca. 4000 B.C. Later, Pepi I (ca. 2300-2180 B.C.) constructed a series of short canals through the First Cataract of the Upper Nile; an extant inscription (discovered in 1906) includes the words, "three cargo boats and four towboats of acacia wood," attesting to the probably navigation of those canals.

Necho, an Assyrian in Egypt ca. 600 B.C. attempted to build a canal from the River Nile to the Red Sea, but he failed to complete it; the lives of 120,000 Egyptians were lost in the process. During the excavation of the Suez Canal by DeLessups in 1866, archaeologists confirmed that King Darius of Persia (521-385 B.C.) completed Necho's Canal; the fragments of a red granite tablet described the opening of the canal in Persian, Median, Assyrian, and Egyptian languages. In a later restoration Ptolemy II ca. 285 B.C. constructed locks with movable gates; if so, that construction predated by many centuries the "invention" of the lock with vertical movable gates in Bruges (present-day Belgium) in the fourteenth century.

There were many Sumerian canals in the valleys of the Tigris and Euphrates rivers: the earliest surviving canal map (on a clay tablet) shows two great canals flowing through the city of Nippur on the Euphrates, and the earliest Sumerian relief (the Stele of Vultures) shows the victory of Lagash in a war fought over a canal in Mesopotamia, but neither of those artifacts is dated. The earliest known date of a canal in that region is the canal built by King Up Nina ca. 2900 B.C. Another canal, the Shatt-el-Hai, built ca. 2200 B.C., carried water from the Tigris to the Euphrates at Ur.

Many of the laws of Hammurabi (ca. 1800 B.C.) were concerned with the details of canal operations such as water rights, the opening and closing of canals, and the use of sluice gates. King Sennacherib (ca. 705-681 B.C.) built a 50-mile-long, stone-lined canal from a mountain spring at Bavian to his capital of Ninevah, Assyria. That canal (excavated in this century) was an advance in canal construction because of the use of aqueducts. An inscription reads, "I caused a canal to be dug to the meadows of Ninevah. I spanned a bridge of white stone blocks. These waters I caused to pass over it." That aqueduct, at Jerwan, was nearly 1,000 feet long and was 39 feet wide. The canal took thirteen years to build.

There is not much known about the construction methods of the early canals. Neither stone nor timber was readily available in the valleys of the Tigris and Euphrates rivers. Rush mats impregnated

Red Flag Canal in China, "under a bridge and through a dam." (Courtesy W. E. Trout III.)

with the natural asphalt that oozed from the ground could have served as the waterproofing needed in the loose soils of the region much as clay "puddling" was used later in the European and American canals. Herodotus described the use of bricks, bitumen as mortar, and the use of rush mats in the building of the great moat around Babylon; it is likely that similar materials were used in the construction of canals. As only the one example of locks has been found (the ones mentioned in Egypt), it is likely that most canals were on one level as is the Suez Canal.

CHINA

Canal construction took place very early in China, independently of that in the Middle East. Though archaeologists and historians have not pinpointed the dates of the earliest canals, they do know that construction was taking place from the Wei River Valley in the North to the Han River Valley in the South ca. 100 B.C. The Pien Canal, the longest section of the longest canal, the Grand Canal, was probably constructed around A.D. 609. Kuo Chou-king, a mathematician under Kublai Khan, completed the last 80-mile section of the Grand Canal from Cambuluc (his capital near Peking) to the Huang River, in A.D. 1293. The Grand Canal at its completion was 650 miles long, with an additional 100 miles of auxiliary canals, reaching from Cambuluc (Peking) in the North to Hangchow in the South.

There is evidence that the Chinese may have been the first to have invented the "pound lock," that is, the conventional chambered lock with gates at each end. Such locks were in use nearly 1,000 years ago accommodating large boats owned by the government. The first pound lock in China was probably invented by Chhiao Wei-Yo on the Grand Canal in A.D. 983. In a later period, the Chinese abandoned the pound-locks in favor of flood-gates with inclined planes or slipways, which they found to be more suitable for the relatively smaller boats then in use.

GREECE

Xerxes, the sucessor of Darius, was also a canal builder. Xerxes built the 1½-mile-long canal on the Isthmus of Mount Athos. In that work he divided the labor among the Egyptians, the Phoenicians and the Greeks, but in the deep cuttings he used only the Phoenicians who had the knowledge of building the canal with sloping sides, in the shape of modern-day canal "prisms." The Greeks had considerable engineering skills, but there is little evidence of extensive canal building other than a canal at Boetia to keep Lake Copais at the desired level, and an attempt by Agamemnon to cut across the Isthmus of Corinth.

ROMAN EMPIRE

The Romans were concerned early with canal building. Some of their accomplishments were the navigable Cloaca Maxima (a great drainage canal that flowed from Rome into the Tiber River) the canal built by Agrippa between Lake Avernus and Lake Lucrinus that was capable of accommodating ships from the Mediterranean; the canal built by Drusus (under Augustus) to convey his army into Germany from the Rhine River to the Issel; a canal attempted by Lucius Verus in Gaul from the Moselle River (near Nancy) to the Rhine; a canal built by Claudius between the Rhine River and the Maese; and, the canal dug by Marius in 102 B.C. from the Lower Rhone River to the Mediterranean.

The Romans approached canal (and other) construction in a methodical manner, using teams of engineers, surveyors, geologists, and inspectors in all parts of the empire. Not all Roman canals were practical, at least not for the technology available at the time, nor completed. Nero had grandiose plans to construct a canal from Rome to Naples, but the project was abandoned at his death. The remains of the canal were discovered in 1507.

Nero also attempted to construct a canal in Greece through the Isthmus of Corinth A.D. ca. 66. He was in good company as Agamemnon, Darius, and Xerxes had all tried before him without success to build such a canal. After several months of excavation, with a half-mile of canal dug in a tremendous cut, by slave labor, Nero abandoned the project. A French company in 1881 chose Nero's route out of four alternatives, but it too abandoned the project. A Greek company with a Hungarian engineer finally finished the canal in 1893. After Nero's death, much of Roman canal building came to a close. There were extensive canal projects in North Africa by the Romans, but there is a lack of evidence as to when they were built and which were completed.

The Arabs took over to a large extent where the Romans left off in the Eastern Mediterranean. The Omayyid caliphs built canals in the Damascus area. Damascus proper had a complex canal system; each of six canals entered the city from a different level of the Abana River in tunnels excavated out of rock.

The Corinth Canal in Greece, a four-mile cut through solid rock to connect the Agean and Ionian Seas. Started by Roman Emperor Nero and finished in the Nineteenth Century. (Photo by the Author.)

EARLY CONTINENTAL EUROPEAN CANALS

The rivers of continental Europe have been navigable to varying degrees for hundreds of years. The Dutch built staunches (flash-locks), a kind of rudimentary device in weirs in rivers which could be opened to allow boats to pass downstream or to be hauled upstream, as early as 1065. The Flemish built staunches or flash-locks as early as 1116, and the Italians as early as 1198. The Flemish may have built a primitive version of the pound lock on the River Reie in the twelfth century, but it was probably some type of a flash lock. (The flash-lock or staunch uses only one gate or a pair of gates at one location, whereas the pound lock uses a gate (or a pair of gates) at each end of a lock chamber.)

The Dutch may have been the first to develop the predecessor of the modern pound lock by a sort of lock at Vreeswijk on the River Leek in 1373. That lock had two gates which encompassed a basin rather than the conventional rectangular lock. The first authentic continental pound-lock (and **perhaps** the first pound lock anywhere) probably is the one built at Dammes (near Bruges in Northwest Belgium) in 1396. Both the lock at Vreeswijk and the lock at Dammes had vertically-rising ("guillotine") gates.

Actually, the canal age on the continent seems to have begun nearly simultaneously in several countries when governmental authorities in the Low Countries and Italy made drainage canals navigable. The Germans built the first canal to have a summit level, that is a canal which had both an ascending and a descending level to overcome. The canal was part of the Stecknitz Navigation which they built between 1391 and 1398 to connect Lake Molln with the River Elbe. It utilized bypass canals, a cut canal, and improved river navigation.

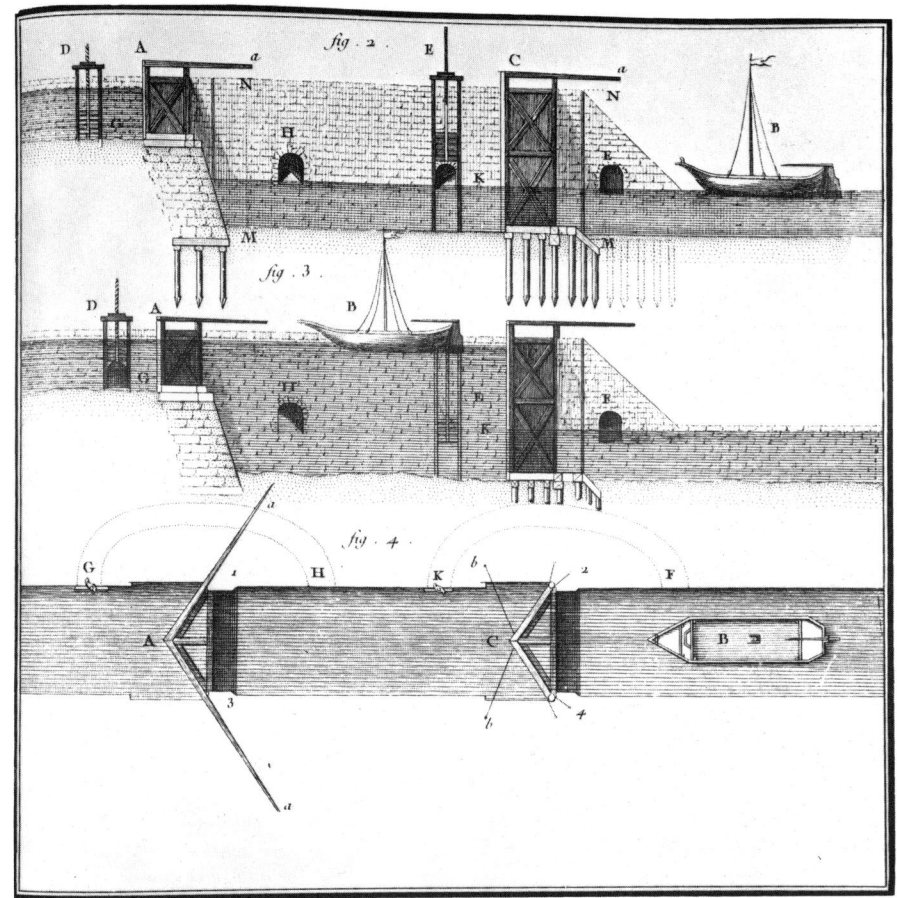

An old print from Europe showing the details of operations of a mitre-gate lock, which remain virtually unchanged today. (Courtesy Bob Mayo.)

Model of the mitre-gate lock developed by Leonardo da Vinci, displayed in the Leonardo Museum, Vinci, Italy. (Photo by the Author.)

Alberti built the first Italian locks near Bologna in 1439. Bertola da Novato under the Duke of Milan built the first Italian canal to overcome a significant gradient, an elevation of 80 feet. That canal was the Bereguardo Canal which he built with pound-locks with mitre-gates (that is, horizontally swinging gates) in Milan ca. 1485. These locks were the first truly modern locks, the principle of which is still used on the majority of the world's inland waterways. It is for that reason that Leonardo is often credited with being the inventor of the canal lock. By 1600, that type gate was in fairly general use as it greatly facilitated the navigation of rivers and canals.

The Briare Canal in France, which connected the River Briare with the River Seine, was the first major continental canal. Henry IV began the canal in 1604 and Louis XIII completed it in 1642. It was 34 miles long and had 41 locks, each 80 feet long and 15 feet wide, very similar to the 90 x 15-foot locks on the Erie Canal, the Chesapeake and Ohio Canal, and many other canals in the United States. The Briare Canal included locks built as staircases, that is, where one lock succeeds the other without intermediate pounds or levels of water. The Briare Canal is still in use today.

The most significant early continental canal was the French Languedoc Canal (later called the "Canal du Midi"), which linked the Atlantic Ocean and the Mediterranean Sea. Francis I first considered the canal and discussed it with Leonardo de Vinci. Leonardo surveyed it, but the canal was simply beyond the technology and the resources of the day. In 1666, Pierre-Paul Riquet began the construction of the Languedoc Canal under Finance Minister Colbert with the general backing of Louis XIV as part of an extensive internal improvement program. The canal opened in 1681, the first canal built to "modern" engineering standards. The Languedoc served as an example and an impetus to canal builders elsewhere on the continent and in Britain, and perhaps even in North America. It was this canal, which set the pattern for the Duke of Bridgewater Canal nearly a century later. The Canal du Midi was nearly 150 miles long, had several aqueducts, a 180-yard tunnel (the first to be built for a navigable canal), a 27-mile-long feeder canal to water the summit level, and 103 locks. In other words, it had all the essential features of a modern, full-fledged navigational canal.

But other major engineering works on canals on the continent did not happen immediately after the completion of the Languedoc Canal, in spite of the good

Inside one of the tremendous locks of the Rhine Canal near Switzerland, with a drop of nearly fifty feet! Note the vertically-rising upper lock gate. (Photo by the Author.)

beginnings of that major undertaking. Several factors were responsible for the delay, among which were the French Revolution, the Napoleonic Wars, new administrations, and the late appearance of the Industrial Revolution on the continent, and particularly in Italy, Germany, and Russia. It was not until about 1815 that the waterways systems of Continental Europe really expanded the building of newly cut-canals and by the improvement of river navigations. These waterways gradually connected industrial areas with the capitals of France, the Netherlands, Belgium, and Russia, and to a lesser extent, other European countries. Most of these improvements were financed by public rather than private means.

EARLY BRITISH CANALS

The history of the British canals begins with the Romans who built the 40-mile-long Caer Dyke from Peterborough on the Nene River to Lincoln on the Witham River, and the Foss Dyke, constructed A.D. ca. 120, which connected Lincoln with the Trent River at Torksey. Henry I improved the Foss Dyke in 1121, and it is still navigable. Unfortunately, the tradition of canal building was not followed in Britain after the departure of the Romans in the early part of the fifth century; the art of canal construction was lost until the seventeenth century. The only inland waterway navigations in use after the departure of the Romans until then were the variously-improved rivers. Some rivers were improved by Edward the Confessor as early as 1065; those included the Rivers Trent, Thames, Severn, and Yorkshire Ouse. By 1724, there were 1,160 miles of improved river navigation in Britain.

John True, ca. 1563-1566, built the first true British canal, a bypass on a section of the River Exe for the Corporation of Exeter. The pound locks that he built there were the first in Britain on either a river or canal. The Exe Canal was 1-3/4 miles long, 16 feet wide, 3 feet deep, and had 3 locks. There is some controversy over whether the locks had vertical gates or mitre gates.

It is as difficult in Britain as in the United States to answer the question, "What was the first canal?" The second question certainly has to be, "In what respect?" For example, the first canal to have industrial significance was probably the Newry Navigation because of the coal it transported in Northern Ireland from Tyrone to Dublin. It was built ca. 1729-1741, was 18 miles long, 45 feet wide, 5 to 6 feet deep, and had 14 locks.

One could argue that the Sankey Canal in Northern England built ca. 1757-1761 was the first "modern" British canal, but most would agree that the canal that ushered in the British Canal Era was the Duke of Bridgewater's Canal in the same general area as the Sankey. This canal was built 1759-1776 to take coal from the Duke's mines at Worsley seven miles to Manchester and soon therafter a branch canal to Runcorn, and is the basis for the main thrust of heavy transport of the Industrial Revolution. One of the distinctive features of the canal was the 200-yard, three-arch aqueduct over the River Irwell. The mines at Worsley had a four-mile main underground canal and 42 miles of side canals at four different levels.

James Brindley built the first extensive canal in Britain, the 93-mile, 74-lock Trent and Mersey Canal that connected the Trent and Mersey rivers by an artificial cut canal; the 2,880-yard Harecastle Tunnel was the first British canal tunnel of any length.

In order to keep construction costs down and to avoid expensive engineering works, the Trent and Mersey Canal and many other early English canals followed the contours of the land, inserting one or more locks where necessary to go to another contour. Those early "narrow" canals had locks 74-feet long by 7-feet wide, accommodating boats in pairs (one behind the other) carrying 25-30 tons of cargo each. (Compare those locks with the standard 90 by 15 locks of the Erie Canal.) The later "broad" canals in Britain were generally about 15 feet wide. Many of the narrow canals are still in existence, providing recreation to thousands of canal enthusiasts. The difference in lock sizes in Britain helped to contribute to the demise of the British canal system for commercial purposes. Interestingly, commercial use of the British canals preceded the canals of the United States by at least thirty years.

There seemed to be little direction in the British canal-building experience from that of continental Europe. Locks, aqueducts, tunnels, embankments, cuttings, inclined planes, and lifts were by native British design for the most. The end of the major canal building era in Britain came about 1835, though navigable inland waterway mileage continued to be added until about 1850 at which time there were over 4,000 miles of navigable inland waterways in England and Wales.

A canal digging device invented by Leonardo in the Museum at Vinci, a few miles east of Florence, Italy. (Photo by the Author.)

The Manchester Ship Canal, officially opened by Queen Victoria in 1894, was the last major canal built in England in the nineteenth century.

With some exceptions, the British canal system was privately built, mainly because there was a pre-existing industry waiting for a better means of transportation for the carrying of coal and other bulky products. On the continent, the canals were largely publicly financed because of the need to encourage the development of industry. In North America, there was a mixture of each.

Canals were essential to the Industrial Revolution in Britain principally because of their ability to accommodate the carrying of bulky commodities at relatively low costs. The most important cargo carried in both Great Britain (and in the United States) was coal; coal to power steam engines, as fuel for industrial kilns and furnaces, to supply gas works, and for domestic heating. As aptly put by canal historian Charles Hadfield, "Coal made the industrial revolution and the need for coal built the canals."

Flight of six locks on the Grand Union Canal at Hatton, England. (Courtesy British Waterways Board)

BRITISH CANAL BUILDERS

The British were a half-century ahead of the United States in building their network of tow-path canals in the "Midlands" and Wales, many of which are still in use today for pleasure-boating. There is no question that the methods of canal construction and operation used by the British influenced early canal design in the United States, with such keen American observers as Elkanah Watson, Canvass White and Robert Fulton carefully examining details of the canals built there, 1770 to 1820. Therefore, it is worthwhile to examine the philosophy of two of the most famous of the British Canal Engineers of the period.

JAMES BRINDLEY 1716 - 1771

James Brindley was an un-tutored Derbyshire millwright whose inventive genius and innovative thinking, plus the sponsorship of the young Duke of Bridgewater, elevated him at middle age to the undisputed position of England's leading Canal Engineer. Brindley is often given credit as the designer of the Bridgewater Canal, the first of the coal canals of England. Actually it was the Duke and his agent, John Gilbert, who conceived of the Bridgewater Canal, with its tunnel drilled directly into the mine diggings at Worsley, to permit the canal barges to load underground and carry their product directly to market at Manchester. It was Bridgewater coal which became one of the key elements in the development of Manchester as the great industrial center of England.

Early in the planning of the Bridgewater Canal, Brindley, already known for his innovations in the milling industry, was hired (1759) by the Duke to assist John Gilbert in the actual construction work. Although frequently at odds with Gilbert, Brindley performed a valuable service in the building of the first canal, and was later given a key role in the extension of the Bridgewater Canal into Liverpool and was also much involved in the building of the first English Canal Aqueduct at Barton.

The word spread of Brindley's technique in overcoming obstacles on the Bridgewater, and his services were sought on numerous other canal projects. Soon he engaged as partners, other persons of similar background to assist in his ever-increasing consulting work on canal design and construction. One of his first major achievements was the design of the 93-mile Trent and Mersey section of the Grand Trunk Canal which started building in 1766, and which, when completed in 1977, with its five tunnels, provided an inland connection between

An old print of Brindley's Barton Aqueduct of 1761, the first structure of its kind in England. (Courtesy Bob Mayo.)

James Brindley (National Portrait Gallery, London)

Brindley traveled constantly, surveying, planning, and advising. He was badly overworked, and completely ignored a diabetic condition, which eventually led to his death in 1771 at the early age of 55. He and his partners had by this time designed and supervised work on 350 miles of canal projects in England, some of which were not completed until after his death. The 8600-foot long Harecastle Tunnel, the Bingley "Staircase of Locks" and other unusual feats of engineering remain as monuments to his inventive genius.

He married, late in life (1765), Anne Henshall, a girl of eighteen. Their marriage produced two daughters only. The "James Brindley", who worked on the old Susquehanna Canal in Maryland, the Schuylkill Canal in Pennsylvania and the Patomack Canal in Virginia may have been a nephew.

THOMAS TELFORD 1757 - 1834

Thomas Telford was born in a remote valley near Langholm in Eskdale, Dumfriesshire, Scotland. He became a working stonemason, first in Langholm, where many of his handiworks are still to be found; then in Edinburgh and finally in London. In the latter city he was "discovered" by William Pulteney, one of the wealthy commoners of England, who became his patron. Through Pulteney, Telford secured a job as a surveyor of Shopshire with the responsibility for all public works in that county. Bridgebuilding became one of his early specialties, and aqueduct design and construction led him naturally into canals.

the Port of Liverpool, on the west coast of England, and the Port of Hull on the east coast.

His fame as a Canal Engineer soon involved him in canal projects throughout the English Midlands, including a connection to the Trent and Mersey, tying in the port of Bristol to the rest of his canal system. Brindley was a great respecter of water power. One of his sayings was that "water is like a giant — safe only when laid on its back". For economy of canal design he preferred that his canals follow the contours of the hills and valleys through which they traveled, creating long "levels". He used locks as sparingly as possible and liked to group them where he could. He was a worrier, and frequently went to bed, night or day - to rest, to think out his problems, frequently coming up with a complete project design before he arose again.

An Aqueduct on the Göta Canal in Sweden, crossing a highway.

Telford was a man of all trades, highly intelligent and with a gift for leadership and the selection of the right men to follow through on his various construction jobs. Unlike Brindley, who was essentially a "man of the earth," working with water and soil – Telford delighted in soaring arches, intricate yet artistic bridges, and the inter-play of metals and masonry. Where Brindley's canals wound endlessly around the contours of the hills, Telford's canals often cut boldly, in a nearly straight line, across country to his objective.

He was at once an Engineer and an artist in the early use of iron as one of his basic building materials. He was a pioneer in the building of eye-bar suspension bridges. In Shopshire alone he built over forty bridges, some of stone, some of cast iron, or a combination of iron and stone -- all of them beautiful, and many still standing today!

An ironmaster friend of Telford's, Abraham Darby, steered him into the position of Engineer and Architect for the Ellesmere Canal Company, whose original goal was to join three rivers -- the Severn, Dee and Mersey. However, the Ellesmere Canal finally became merely a series of navigations, proceeding from the River Dee in the vale of Llangollen, a distance of 112 miles, and including the Chester Canal. A major project on the route was the crossing of the valley of the Ceiriog at Chirk, which Telford spanned with a masonry aqueduct, having ten spans of forty feet each, carrying the canal seventy feet above the water.

However, the nearby Pontcysyllte Aqueduct was Thomas Telford's masterpiece, and one of the most notable canal structures of all time! This aqueduct, (still standing) carries the Llangollen Branch Canal over the River Dee at an elevation of 127 feet. It is 1007 feet long,

The Pontcysyllte Aqueduct on the Llangollen Branch Canal, crossing the River Dee at an elevation of 127 feet. Designed by Thomas Telford, 1795.

with nineteen arches, and a cast iron trough. Built 1795 to 1805, all appurtenances of this amazing structure remain in almost perfect condition to this day!

Telford continued to build canals in England, Scotland and Sweden for the rest of his lifetime. His next major project was the building of the Caledonia (Ship) Canal to connect the Atlantic Ocean with the North Sea -- across Scotland -- a tremendous undertaking. It was a work worthy of Telford, who started the actual construction in 1804, assisted by a capable resident Engineer, Mathew Davidson. After many difficulties and delays, the canal was finally opened to traffic in 1847, with 28 locks to allow ocean-going vessels to surmount the 102-foot summit level at Loch Oich. The Canal was never a financial success, as boats had grown to exceed the width for which the locks had been designed and somehow, towing paths had been forgotten. It had been planned originally as a military route to avoid travel around the tip of Scotland when enemy ships threatened the North

Thomas Telford ("Thomas Telford" — Bracegirdle & Miles)

Sea. In this connection, it was used successfully in both the First and Second World Wars.

Telford had better luck when he went to Sweden to supervise the construction of the Göta Canal from the North Sea to the Baltic. The Swedes were excellent canal builders, having built a canal as early as 1606, with locks, between Eskilstuna and Lake Malaren, but they bowed to Telford's superior experience on the Göta.

Telford's last canal was the Birmingham and Liverpool Junction Canal to up-date the previous canal, which had 99 locks in 114 miles. Telford presented a plan to reduce the length of the canal by 20 miles and the number of locks by thirty. The plan was accepted and work began, but the new canal was still under construction at his death, in 1834. Thomas Telford remained a bachelor, never seeming to find time to become interested in the opposite sex.

A traditional "Narrow Boat" rising in a lock on the upper Thames in England. Built to fit the narrow locks on many of the English canals, two of these boats can be locked through the conventional size locks, side by side. (Photo by the Author.)

THE VISIONARIES

Inland waterways and canals in America had been dreamed of and talked about a full century before the first true canal was dug here. William Penn, having laid out Philadelphia in 1682 (which became the new country's first great seaport) issued proposals for another inland city in Pennsylvania (1690) as follows:

"It is now my purpose to make another settlement upon the river Susquehanagh . . . and the most convenient place for communications with former plantations in the east . . . which will not be hard to do by water by benefit of the river Scoulkill, for a branch of that river (Tulpehocken Creek) lies near a branch that runs in the Susquehanagh River (Swatara Creek) and is the common course of the Indians with their skins and furrs into our parts . . . from the west and northwest parts of the continent."

This canal connection was discussed for many years, and in 1762 David Rittenhouse, the astronomer, and Dr. William Smith, provost of the University of Pennsylvania, made surveys over Penn's route from Reading to Middletown, the route later traversed by the Union Canal.

For years, Benjamin Franklin had yearned for canals in America, even before the Revolution. In 1772 he wrote the Mayor of Philadelphia: "Rivers are ungovernable things, especially in hilly countries. Canals are quiet and very manageable." He then proceeded to outline a complete plan of his own for a canal system, based upon a Susquehanna-Schuylkill connection, which would bring wealth to Philadelphia from the hinterland.

Thomas Jefferson, while United States Ambassador to France in 1787, took an eight-day trip along the Canal Du Midi in France which he said he had a "great desire to examine minutely, as at some future time it may enable me to give information thereon to such of our States as are engaged in works of that kind". Jefferson's papers indicate that he made sketches of the lock operations and a sketch for improvement of the wicket gates, which were operated by a time-consuming screw arrangement. Jefferson sent a report of his findings to President Washington. There is evidence also that at some time in his early career he talked with Alexander von Humboldt about future canals across the Isthmus of Panama!

Robert Fulton - Self Portrait (New York Historical Society)

ROBERT FULTON
1765 - 1815

Best known for his Steamboat "Clermont", successfully tried out on the Hudson River in 1807, Robert Fulton was, among other things, interested in the English Canals and invented some devices for use on canals for small boats.

Born in Lancaster County, Pennsylvania in 1765, Robert Fulton was a genial lad who early exhibited artistic talent and began his career, at the age of seventeen, as an artist in Philadelphia. Endowed with an outgoing nature and great personal charm, Fulton quickly built up a clientele of some of Philadelphia's leading citizens, including the influential Benjamin Franklin. Taken by the young artist's talents, Franklin sent Fulton to London in 1786, with a personal letter of recommendation, to study under another American artist who had established an excellent reputation in England -- Benjamin West.

Young Fulton made a favorable impression immediately on West, who took him into his household; it was the beginning of a life-time friendship. West was of great help to Fulton in introducing him into the art circles of Europe. In 1789, Fulton wrote home to his mother that he had been working hard at his paintings since arriving at West's home, and that two of his canvases were already on display at the Royal Academy in London. Shortly after this he began

Drawing by Robert Fulton, showing one of his ideas for lifting small canal boats from one level to another. In 1794 he obtained Patent #1988 from the British government for an inclined plane system similar to those now in use on the French Canals. (Courtesy New York Historical Society, Randall J. LeBoeuf, Jr. Collection.)

traveling around England, making copies of some of the art treasures of the British nobility, supporting himself entirely on his own artwork. While so engaged, he had the good fortune to meet the Duke of Bridgewater and the Earl of Stanhope, and he became engrossed in canal-building with the former, and steam navigation with the latter. It was these two contacts which changed Fulton's interest from Art to Engineering.

His contact with the Duke of Bridgewater led him to spend about eighteen months at work on a canal being built in the vicinity of Birmingham. While there he made the acquaintance of James Watt who had recently perfected his new steam engine. Fulton made a careful study of Watt's invention, and later ordered a Watt engine for the "Clermont".

After his experience working on the English canals he published a series of canal essays in the London "Morning Star", and in 1794 obtained from the British Government a patent for a double-inclined-plane tank system for raising canal boats from one level to another, without the use of locks. He also invented several canal-dredging machines. His skill as an artist resulted in many excellent sketches of his inventions, easily comprehended by the layman. In 1796 he published a "Treatise on Canal Navigation" sending copies to both George Washington, then President of the United States, and also to Governor Mifflin of Pennsylvania, urging their consideration of the great advantages which would accrue from the use of Canals in America.

Robert Fulton's contact with the Earl of Stanhope, who had already been experimenting with the idea of a steam-driven vessel, led to further discussions with other contemporary English and American inventors and ultimately to a partnership in 1802 with the influential and wealthy Chancellor Robert Livingston, American Ambassador to France who for years had been interested in steam navigation. Out of this partnership grew the successful "Clermont", which steamed up the Hudson River in 1807 to the amazement of thousands of observers. Settling permanently in New York, Fulton married Harriet Livingston in 1808 and had a family of three daughters and a son. Other inventions by Fulton were: the World's first steam Battleship (1814); the World's first Submarine, (intended for Napoleon) in 1800. In his latter years he also submitted plans to the Governor of New York for a canal between Albany and Lake Ontario. He died in 1815.

ELKANAH WATSON 1758 - 1842

Another young visionary, who deserves special attention, was Elkanah Watson. Not much is known about the personal history of this man, but his name is found in accounts of early canal developments in America. For instance an article published in London in 1803, refers parties interested in American canals to the "Journal of Mr. Elkanah Watson, a gentleman who has traveled much both in America and Europe."

Elkanah Watson was a young New Yorker who was an emissary, during the Revolution, from Continental Congress to Benjamin Franklin, while the latter was in Paris to woo the French to the American cause. Watson remained in Europe for five years (1779-1784) during which time he examined, with great attention to detail, the canals of France, Belgium, Holland and England. After his return to the United States, he visited George Washington (1785) and discussed with him a canal connection between the Potomac and the Ohio River. For the next few years he traveled through the American interior investigating the possibilities of canals at various points. He kept a written record of his findings, which later became known as the Watson "Journal of Travels".

In 1788, while exploring the upper Mohawk River in New York State, he wrote: "In contemplating the situation at Fort Stanwix, at the head of the Batteaux navigation on the Mohawk River . . . . I am led to think this station will in time become an emporium of commerce to Albany and the vast western world above. Should the Little Falls ever be locked . . . and other obstructions removed . . . to Fort Oswego, who can reasonably doubt . . . that the State of New York have it within their power, by a grand stroke of policy, to divert the full trade of Lake Ontario, and the Great Lakes above to Albany and New York." Watson was named by the New York legislature, in 1792, as a member of a Commission to explore a route which would connect the Hudson with the Great Lakes. Thus was laid the groundwork for the Erie Canal.

GEORGE WASHINGTON
1732 - 1799

By Robert S. Mayo, P.E.

We all know of Washington as a wealthy planter, as a victorious general and as our first president. But few people realize that he was also a distinguished engineer and that his success in other fields, particularly in the military, were due in no small part to his engineering training and experience. There were several highways and canals in which Washington played either a direct or an indirect part. The highways are still in daily use and you have probably traveled over them. Of the canals only one is still in use.

In 1747 Washington was employed by Lord Fairfax as an apprentice surveyor in the wilderness south of the Potomac River, just east of what is now Cumberland, Maryland. This sixteen year old boy was faced by all the hardships of the backwoods country but in these three years he learned surveying and the way of the Indians. What he learned about the Indians was most valuable in the French and Indian wars. At the age of nineteen he was examined and licensed as a state surveyor of the Virginia colony.

At the urging of his brother Lawrence he returned to the big house along the Potomac to complete his formal education. He was tutored by friends of Lawrence. They must have been excellent tutors and George an apt scholar. He was taught the Social Graces, the outward signs of a gentleman. He was tutored in Military Science and Tactics and it was Washington's skill in this field that won the Revolutionary War for us. And he was tutored in Engineering.

George Washington, 1795 (From the original portrait in the Boston Athenceum)

Some of the difficulties which beset the builders of the first Dismal Swamp Canal. The ground was so swampy it was impossible to build the ordinary tow-path. The boats were first shoved along by men walking on foot logs. (From Harper's Magazine, 1856; courtesy Bob Mayo.)

The latter was primarily Military Engineering because there was at that time no clear-cut differentiation between civil and military engineering. It was to be another fifty years before the first college of engineering opened its doors in North America. This was West Point, established on the recommendation of President Washington.

George Washington was sent in late 1753 by the Colonial Governor of Virginia when it was heard that the French were moving troops into the Ohio River Valley, then part of the Virginia colony, with a message to the French Commander to "cease and desist." Washington was courteously received by the French Commander at Presque Isle, but told that they had no intention of leaving.

Washington made many observations of French activities and west Pennsylvania terrain during this trip which were of great value to the British during the course of the ensuing war. He subsequently led a force of Virginia militia, in 1754, into Western Pennsylvania, which was forced to surrender when besieged at Fort Necessity by superior French forces.

Major Washington was an Engineer on Braddock's ill-fated expedition of 1755

George Washington is shown here as a young surveyor, with magnetic transit in hand, and at his feet a surveyor's chain. (Library of Congress print.)

to capture the French fort at the Forks of the Ohio River. From Fort Cumberland he helped locate a road westward and northwest to what is now known as Pittsburgh. In the western part of the state they still call it Braddock's Road but we know portions of it better as Route US-40, or the "National Road."

In 1758 he was a Lt. Col. on the staff of General Forbes. This expedition was successful in capturing Fort Duquesne, the name used by the French for their Fort at Pittsburgh. General Forbes located a new road from Fort Bedford in central Pennsylvania to Pittsburgh which is still known as Forbes Road, but we know it better as Route US-30; or the Lincoln Highway.

During these various military expeditions into the western territory, Washington observed the potential of the Potomac River for an improved western water route, at least to the foothills of the Allegheny Mountains. He made recommendations on the matter to the Virginia legislature in 1774, but was unable to get action.

For 8-1/2 years Washington was Commander-in-Chief of the American Revolutionary War Forces. At the end of the war, in 1784, he took a trip from Mount Vernon up the Potomac River, over the Allegheny Mountains and down the Monongahela to the infant city of Pittsburgh. Then he continued by boat down the Ohio River, up the Kanawha River and crossed the mountains close to White Sulphur Springs.

He saw on this trip that unless communications were improved the settlers in the rich Ohio Country would ship their products and buy their supplies from New Orleans, which was then in the hands of the Spanish. As he put it, "The Western states hang upon a pivot; the touch of a feather will turn them any way".

Washington was a subscriber to all the scientific journals published in France or England. He had read there of the wonderful success of the canals as built in those countries. Also, he was kept informed by such respected personalities as Elkanah Watson and Thomas Jefferson of canal activities overseas. In the years following his western trip he was an incorporator, a stockholder and sometimes an officer of several canal companies designed to improve communication between Chesapeake Bay to the west or south.

The "Patowmack Canal" by-passed the Little Falls and Great Falls not far above what is now the city of Washington. This "Pawtomack Co." was incorporated in 1774 but work did not begin until after the war. Washington was elected President of the company in 1785.

Of considerable interest is the fact that Washington chose James Rumsey, of Shepherdstown, Virginia (later West Virginia) to be his chief engineer. Rumsey is best remembered for his water-jet propelled steam boat on the Potomac, which contested John Fitch's steam boat as the first in the United States. Rumsey stayed on for a year, but had so many difficulties with subordinates and run-away slaves, that he quit the job in July of 1786, to renew his attention to his patent fight with John Fitch.

The Patowmack Canal was finally opened in 1802 but was closed in 1830 when the C & O Canal was opened. A visitor to Great Falls, on the Virginia side, can still see the remains of this ancient canal, one of the first built in North America.

George Washington was also asked to be President of the James River Company, formed in 1785, but declined. This company was supposed to build locks around Richmond and open the upper portions of the James River. In later years, long after Washington's death, it was reorganized as the James River & Kanawha but work was not really underway until 1835. It was extended to the foothills of the Allegheny Mountains but it was never completed and was finally taken over by the Railroad in 1880.

The Dismal Swamp Canal connects Chesapeake Bay with Albemarle Sound. Washington, in 1763 made the first reconnaissance of the proposed route in a canoe. Work was started in 1793, the first boat went through the uncompleted canal in 1805 and it was fully opened in 1812. It proved invaluable in transporting supplies during the War of 1812 and it is one of the few old canals still in daily use as part of the Intracoastal Waterway.

As President of the United States, Washington never lost interest in canals and while visiting Reading, Pennsylvania, in November 1783, he rode out to Womelsdorf, 14 miles away, to inspect the locks of the Susquehanna & Schuylkill Canal, then building. Later it was re-named the Union Canal.

EARLY CANAL BUILDERS IN AMERICA

Our young nation, beset by political problems and a general lack of funds in the latter days of the Eighteenth Century, nevertheless at the urging of President Washington, did make a start at building tow-path canals and making river improvements. Most of them were very short connections around rapids or other impediments in the rivers of the Northeast.

The Schuylkill and Susquehanna Canal (later known as the Union Canal) completed the first 15 miles of work, east of Lebanon, Pennsylvania (1792-94) but this note-worthy beginning completely exhausted their funds. Work was not resumed for another 27 years!

The original Dismal Swamp Canal was opened for small boats in 1784, and enlarged in 1807 to permit six-foot 'flats" to pass through. It was finally opened to full navigation in 1812.

To the south a short canal was opened in 1785 from the Spanish controlled Port of New Orleans to make a more direct connection east, as part of the coastal route to Florida. It was known as the Carondelet Canal. This was acquired by the United States as part of the Louisiana Purchase in 1803.

A significant improvement was made in New York State, 1783-95, known as the Little Falls Canal, on the Mohawk River, later to be superceded by the famous Erie Canal.

Seal of the "Proprietors of Locks and Canals" showing the inclined plane used at South Hadley, Massachusetts for passing canal boats between high and low levels.

The two-and-a-half mile South Hadley Canal around a dam on the Connecticut River in Massachusetts (1792-95) employed a unique device to overcome the change of levels, known as the South Hadley Inclined Plane. The first successful inclined plane anywhere had been installed on the Ketley Canal in England in 1788 but when Engineer Benjamin Prescott decided to use the inclined plane at South Hadley, he had no precedent in this country to guide him. It is a tribute to American engineering inventiveness that this plane worked successfully until 1848.

The James River Company, completed their first seven-mile canal from the basin at Richmond to the head of the Falls at Westham in 1789, which permitted small boats to pass the Falls. It was not until further enlargement was completed, in 1795, that the first freight boats were able to get through.

The Conewago Canal, a one-mile by-pass around the York Haven falls on the lower Susquehanna in Pennsylvania was completed 1792-97, allowing two-way traffic for the first time at this point, via locks.

The Patowmack Canal at Great Falls, Virginia was opened to traffic in 1802, and subsequent improvements were made, using a number of short by-pass canals around river rapids of the Potomac and Shenandoah Rivers to Harpers Ferry, and as far west as Sawmill Falls.

The Susquehanna Canal, on the east bank of that river in Maryland was incorporated in 1783 and opened to partial traffic in 1802.

In this connection, Benjamin Henry Latrobe was an Engineer who ran surveys for improvement of the Susquehanna, Columbia to Tidewater, 1801-1802.

However, the first canals in America of major significance, were the 22-mile Santee and Cooper Canal (1793 - 1800) in South Carolina, and the 27-1/2 mile Middlesex Canal, connecting Lowell and Boston, Massachusetts, completed 1794-1804.

JOHN CHRISTIAN SENF
1753 - 1806

The first true canal in America was built to connect the Santee and Cooper Rivers in South Carolina, so that the Santee, which served a good portion of the interior of both North and South Carolina, could be connected with the Port of Charleston, via the Cooper River. Many of the goods produced in the interior -- primarily Indigo, rice and (later) cotton, frequently were lost in navigation of the treacherous, swampy lower Santee, and the even more dangerous sea voyage from the mouth of the Santee to Charleston. In 1786 the South Carolina Legislature signed a charter for the construction of a tow-path canal to connect the Santee and Cooper at the closest point above Charleston, but work did not begin until 1793.

Supervision of the work was entrusted to Colonel John Christian Senf. He was a Swede who had come to America with Burgoyne's Hessian troops and fallen into the hands of the Americans with the surrender at Saratoga.

During the latter years of the War, Senf served as an engineer with the South Carolina militia. Afterward, he became Chief Engineer for the State of South Carolina. He was a master craftsman of great energy and vast pride who insisted on doing things his way -- a vain and jealous man who often permitted his professional judgment to be influenced by personal obligations -- to the detriment of the work at hand.

Nevertheless the work on the Santee Canal proceeded under his guidance, surmounting unbelievable difficulties. Most of the work was done by slaves who were frequently needed for "more important" work on the plantations. The white workmen died like flies in the feverish summer-times. One of the most serious problems was the opposition of the local landowners, through whose property the canal passed. Lawsuits were frequent.

Senf favored certain property owners, with the result that the canal was finally located so that its upper reaches had to be fed by artificial reservoirs, instead of nearby streams. The proprietors of the operation would have been glad to replace Senf, but no other qualified engineers were available. Senf did much of the detailed supervisory work over the entire 22-mile route of the canal personally.

In 1800 the canal was finished -- a tremendous accomplishment. Where, heretofore, canals in America had been simply short by-pass channels around river falls, the Santee was the first true, full-fledged canal, cutting boldly across country. A local historian, F. A. Porcher, wrote: "The Canal is 22 miles in length, 35 feet wide at the surface, 20 feet at the bottom, depth 5-1/2 feet, with four feet of water, capable of carrying boats of 22-tons burden. On each side is a draw-path ten feet wide. It has two double and eight single locks, and in its course over the country it lies over eight aqueducts or culverts through which as many swamp streams find a passage under its beds. From the Santee it rises by locks 34 feet to the summit level, whence it descends by seven locks to Cooper River 69 feet, making the difference of level between the two rivers 35 feet".

During its period of operation, 1800 to 1850, it was a monument to the grim tenacity of John Christian Senf, who with all his faults, had pushed it through to completion, where lesser men would have failed in the very early stages.

LOAMMI BALDWIN
1745 - 1807

The second true canal in America was the Middlesex, built to make an inland connection between the Charles River at Boston and the Merrimack River, near the present location of Lowell, Massachusetts. The objective of the canal was to provide an inland route for lumber and other important materials being shipped from New Hampshire to Boston. It was later a key factor in the rapid growth of Lowell as one of the first great textile centers in the United States.

Loammi Baldwin was born the son of a carpenter in Woburn, Massachusetts - - his later permanent home. As a young lad he was apprenticed to a cabinet-maker, and later attended weekly lectures on mathematics and physics at Harvard. In 1765 he became a full-fledged land surveyor and engineer. He served as a Major with the American Revolutionary forces and saw action in the battles of Lexington, Long Island and Trenton in 1776. Leaving the service, he continued in the practice of civil engineering, serving also in various public offices in Woburn and Middlesex county. Along with James Sullivan, he was an ardent promoter of the frequently-discussed Middlesex Canal. John Hancock, then Governor of Massachusetts, signed (1793) the document "Incorporating James Sullivan, Esq., &

Loammi Baldwin (Middlesex Canal Association Archives)

The Middlesex Canal, passing Loammi Baldwin's home in Woburn, Massachusetts. (Middlesex Canal Association Collection.)

others . . . as Proprietors of the Middlesex Canal". Even though Baldwin protested his lack of knowledge, he was named in 1794 as Chief Engineer and Superintendent of the project.

Baldwin immediately began a search for a qualified canal engineer in the New England area, to assist him, and discovered that no such personage existed. However, he had heard of the activities of William Weston, an English canal engineer, who was then a consultant of the Schuylkill and Susquehanna Canal in Pennsylvania and made a trip to Philadelphia to meet Weston, finally persuading the latter to spend some time at Boston, running surveys for the Middlesex. He had hoped to induce Weston to continue work on the Middlesex, after the original surveys were complete, but was unsuccessful. Weston returned to Philadelphia, and later to England, subsequently refusing the job of Chief Engineer on the Erie Canal.

William Weston ran two possible route surveys and made a detailed report to the Middlesex Proprietors. Traveling with Weston, Baldwin was able to learn a great deal from the Englishman, enabling him to proceed with the design and construction of the Middlesex. It was a monumental undertaking indeed for an engineer with no prior experience with hydraulics, lock operation or other fine points of canal design. No doubt there was considerable correspondence between Baldwin and Weston, while the latter remained in America.

Baldwin showed his ingenuity and expertise as one of America's first self-taught canal engineers, and pursued the Middlesex Canal construction with great devotion and energy until the canal's opening, late in 1803. His son, Loammi Baldwin II, (1780-1838) followed in his father's footsteps and became Chief Engineer of the Union Canal, connecting Middletown and Reading, Pennsylvania, and was later selected as the first Principal Engineer of Virginia.

The Middlesex, when completed, extended 27-1/2 miles from Boston to Lowell, having 20 locks, 7 aqueducts, and 50 bridges. In full operation until 1853, it became a field-study project for many of the engineers on the Erie Canal, and an example of early American engineering at its finest.

WILLIAM WESTON
1753 - 1833

In the early days of canal-building in America, there were a few European engineers who made a considerable contribution. Chief among these was William Weston, born in central England at the beginning of the great canal-building era under James Brindley. It is thought that young Weston may have been an apprentice to Brindley at one time. In any event, Weston had great admiration for Brindley's works and a keen desire to become an engineer himself. One of his notable engineering works in England (about 1786) was a three-span stone-arch turnpike bridge over the Trent, at Gainsborough.

Hearing of Weston, the proprietors of the Schuylkill and Susquehanna Navigation Company (later the Union Canal) wrote him (1792) and persuaded him to come to America to act as their consultant. Weston arrived in Philadelphia with his bride, in January of 1793, and went to work immediately. He brought with him the Troughton "Wye Level", unknown in America, which was soon being put to use on almost every canal project in the United States.

Loammi Baldwin, commissioned to build the Middlesex Canal in New England, sought out William Weston, who in a few weeks got things started on this canal to connect Lowell and Boston. When the Schuylkill and Susquehanna ground to a halt, due to lack of funds, Weston was engaged by the builders of the Philadelphia-Lancaster Turnpike (1792-95), first hard-surface, inter-city highway in the United States. Weston's reputation spread, and in the latter days of the construction of the Potomack Canal, he was called by company President George Washington to assist in solving some of the problems which beset them at Great Falls, Virginia.

He was then retained by the State of New York (1794) to run surveys for various improvements on the Mohawk River in the Fort Stanwix area. While working there, Weston hired young Benjamin Wright to assist him.

Weston's work took him from upper New York State to Virginia, to Philadelphia and to New York City. Demand for his services kept him continually on the move, and the tiring, primitive travel conditions which existed in the hinterlands of America in those days may have influenced his early decision to return to England, about 1800.

Before leaving, however, he completed a few more engineering projects, such as a design for cofferdams for the piers of Timothy Palmer's "Permanent Bridge" across the Schuylkill in Philadelphia, and recommendations for a water supply for New York City -- tapping the Bronx River to the North to replace the wells and ponds in lower Manhattan.

In 1811 he was asked to review plans (by mail) for the Erie Canal, and in 1813 was offered the job of Chief Engineer of the Erie Canal, at a figure which the Commissioners were sure would bring him out of retirement. Reluctantly, Weston refused and his "pupil" -- Benjamin Wright, was chosen instead.

THE ERIE CANAL

De Witt Clinton ("Old Towpaths" — Harlow)

At the turn of the Nineteenth Century, construction of tow-path canals in America languished and virtually ceased, for the next 15 years. Except for extensive sales of lottery tickets for the Union Canal, and the Albert Gallatin Report of 1808, there was little to maintain public interest in canal building. Money, too, was lacking and the War of 1812 with England diverted attention for several years from internal improvements.

Only in New York State was interest in canals kept alive -- first by such interested parties as Elkanah Watson, who conducted a personal public relations campaign for a water connection between Albany and Lake Ontario; later by General Schuyler and William Weston, who in 1797 made an exploratory tour of upper New York; by James Hawley, a prominent New York Citizen who wrote a series of newspaper articles (1807) on the value to the city of a navigable waterway between the Hudson and Lake Erie; and -- finally -- the redoubtable DeWitt Clinton, mayor of New York City and later State Governor.

A delegation of two New York State legislators, Judge Forman and William Kirkpatrick, had approached President Thomas Jefferson in Washington, after the Gallatin Report had been made public, seeking some of the $20 million mentioned in the report. Jefferson, who had just signed a bill (1806) to get work started on the famous National Road was cold to the entire idea of the Erie Canal. "It is a splendid project" he said "and may be executed a century hence . . . . here is a canal of a few miles, projected by General Washington (The Potomack Canal) which has languished for many years because the small sum of $200,000 . . . cannot be obtained. And you talk of making a canal three hundred and fifty miles long through a wilderness! It is little short of madness to think about it!"

But there were many prominent citizens of New York who would not give up on the idea. About 1808, the New York legislature appropriated $600 for a survey, which Judge James Geddes, of Onondaga County, was asked to run. Geddes, a lawyer, with some experience in surveying, ran surveys from Albany to both Lake Ontario and Lake Erie, using the natural waterways as much as possible. When he made his report a year later, Geddes argued that if the Lake Ontario connection were used, much of the trade might be turned aside into the St. Lawrence River, whereas a direct connection with Lake Erie would assure through trade with the western states in the USA. The War of 1812, and the threat of further hostilities to the North, crystallized the determination of the Legislature to pursue an all-American route to Lake Erie.

There was considerable opposition to the Erie Canal, particularly by representatives from New York City, since it now appeared that the full expense of such a project would have to be borne by the State, with no help from the federal government. DeWitt Clinton, the man later to be called "the builder of the Erie Canal" had at first expressed little interest in the project. His friends won him over, and he soon became one of the most vigorous promoters of the Canal. At the close of the War with England, a group of responsible citizens, headed by DeWitt Clinton, met in New York City (1815) and signed a petition explaining the benefits of the proposed canal, which was circulated throughout the State, particularly to those counties through which the canal would pass. As a result the State Legislature, at its next session, received appeals from more than 100,000 of its constituents to get the canal started! They subsequently appropriated $20,000 for detailed surveys. In the meantime, DeWitt Clinton, who had served previously as Mayor of New York, was elected (1817) Governor of the State, putting him in the most favorable position to see the project through. He predicted that the Canal would be finished in ten years, and was re-elected Governor in 1821, as it became obvious that his ten-year prediction was conservative.

It was decided that not only would the Canal be built across northern New York State, but that a separate Canal would also be dug to connect the Tidewater Hudson (near Albany) with Lake Champlain, in anticipation of a connection into the St. Lawrence via the Richelieu River. For the final survey work, Benjamin Wright was assigned to the Erie, and James Geddes to the Champlain. While these two men had both had some surveying experience, neither of them could lay claim to the title of Engineer. Yet the two of them assumed the technical direction of the biggest engineering job yet at-

Governor De Witt Clinton dumps a keg of Lake Erie water into New York Harbor. This ended the triumphal journey of a flotilla of boats from Buffalo to New York City at the opening of the Erie Canal, (1825). (Culver picture; Courtesy American Legion Magazine)

tempted in America, and with some able assistance, carried it through to completion!

Canvass White, a young engineer with the Wright survey team, turned out to be the real engineering genius of the Erie Canal. He had studied surveying, mathematics, astronomy and other subjects at Fairfield Academy. His work in running levels west of Rome, N. Y., soon came to the attention of Benjamin Wright, and later, DeWitt Clinton himself, who saw rare possibilities in the young man. At Clinton's suggestion, Canvass White was sent to Europe in the fall of 1817 to inspect the canals of the Old World and to obtain some up-to-date surveying equipment. White walked 2000 miles along the canals of Great Britain, studying every feature. He returned the following year with copious notes and drawings, and new instruments.

In the meantime there was great indecision as to whether to build the locks of wood or stone. Proper cement for the stone could be procured only from Europe, and at great expense; wood would be perishable in a few years. Doubtfully, the decision was made to use stone, putting the blocks together with quicklime mortar, and using the expensive European cement only for "pointing." However White, within a few months after his return from abroad, discovered a deposit of stone near Chittenango, on the line of the canal, from which an excellent grade of hydraulic cement could be made!

The actual work of construction began (July 4, 1817) at the center section of the canal, where the going was easiest. By December of 1817, between 2000 and 3000 men were at work and 15 miles of canal had been completed. A number of Yankee innovations had been developed such as the plow and scraper, dumping wheelbarrows, sharp-edged shovel, etc. to speed up the manual work. An unusual tree-cutter, as well as a stump puller were invented on the Erie Canal. Difficulties were experienced in the Montezuma marshes near Syracuse, where the ground was so saturated with water that excavation was postponed till winter when the ground was frozen. In the summer of 1819, a thousand men died in the same area, of malaria. The lessons of the Panama Canal had still to be learned, and the importance of eliminating mosquitoes! In spite of many set-backs, and constant heckling by the opponents of the canal, the work continued with short sections of canal being opened to boat traffic as they were completed. One of the most difficult feats of the entire project was the climbing of the Niagara Encarpment, a solid-rock ridge at the west end of the route. Here Nathan Roberts pierced a solid mixture of limestone and flint, using black blasting powder to cut a flight of five double locks, each with a lift of twelve feet — thus producing the famed Lockport Locks. It took two years to do it, but when completed, every man on the job was proud of every lock that they had blasted out of nearly solid rock. This was the major obstacle to join the canal with the waters of Lake Erie, at Buffalo.

When the canal was opened on October 26, 1825, nearly two years earlier than planned, the populace of the entire State declared a holiday, while De Witt Clinton and a triumphal group of dignitaries officially opened the canal with a flotilla, starting in Buffalo and winding up in New York City, preceded by cannon-fire along the entire route, with a "wedding of the waters" ceremony at the New York City end. De Witt Clinton personally dumped a container of Lake Erie water into New York Harbor, symbolic of the all-water connection to the Great Lakes. It was a great day in the history of the

The "Emita II," only canal packet boat presently offering regular service on the Erie Barge Canal System, emerges from the Mechanicsville, New York lock on the Champlain Canal. (Photo by the Author.)

State of New York, marking the completion of one of the most ambitious engineering achievements of mankind to that date. The Erie Canal also had an immediate effect on the entire history of transportation in the United States, as we shall see.

Almost overlooked in the excitement of the last few months of the building of the Erie, was the opening in 1823 of the Champlain Canal. The architect of this project was James Geddes, assisted by a French Engineer named Marc Isambard Brunel, who had done the original survey work between the Hudson River and Lake Champlain. Construction on the Champlain Canal began in 1818 and it extended sixty-six miles upstream from two entry points: one, a series of locks off the Hudson River in the village of Waterford; and the other, at the Cohoes Junction with the Erie Canal, via the Waterford southern tier. Northern terminus of the canal was Whitehall, on a navigable arm of Lake Champlain. The Lake extends north across the Canadian border, where it becomes the source for the Richlieu River, running north to the St. Lawrence.

After normal relations had been re-established with the Canadians at the close of the War of 1812, the Legislature of Lower Canada passed a bill (1818) granting authority for a canal to by-pass the Chambly Rapids on the Richlieu. A lock and dam had previously been built at St. Ours, to permit slack-water navigation along the Richlieu to the Chambly area. Numerous delays and financial difficulties beset the Chambly Canal, which was initially opened in 1843, but due to poor construction, had to be rebuilt. It was finally opened to through traffic in 1858, 12-miles in length, with 9 locks.

Thus in 1858, a direct water connection was completed between the Port of New york and the Ports of Montreal and Quebec on the tidewater section of the St. Lawrence River — a major achievement for both countries, and the beginning of expanded trade between them. Needless to say, the economy of the Champlain Valley in northern New York State and Vermont was favorably affected by this new, inland route to upper Canada.

Before we leave the Erie and Champlain Canals, we want to examine, in further detail, the personal histories of the four giants of construction of both canals, which had earned them the right to the title: "Canal Engineer."

BENJAMIN WRIGHT 1770 - 1842

Benjamin Wright was born on a farm in Wethersfield, Connecticut. Early in life he was tutored by his uncle, Joseph Allen Wright, in the rudiments of both law and surveying. His family moved to Fort Stanwick (now Rome) New York in 1889 and young Ben found surveying in that area more profitable than farming. In a few years he had built up a substantial practice, which included road and canal surveys, and some two million acres of property in St. Lawrence County. He was appointed County Judge in 1812. His early survey work included assistance to William Weston, the famous English Engineer,

Benjamin Wright (ASCE Biographical Dictionary)

who in 1794 ran surveys in the Little Falls area, for what later became part of the Erie Canal. When the Erie Canal Commission was formed, they first attempted to persuade William Weston to come out of retirement in England as their chief engineer. When Weston refused, on the grounds of advancing age, and at the urging of a local citizen of influence, Joseph Ellicott, they turned to Judge Benjamin Wright. At first, Wright was placed in charge of the center section of the canal only, with James Geddes on the western end, and Charles Broadhead on the eastern section. However, in view of Wright's excellent work on the central section, where the work began, he was soon made Chief Engineer of the entire operation, from the Hudson to Lake Erie.

The Erie Canal became the first great American school of engineering. Judge Wright's best "pupils" were Canvass White, Nathan Roberts, David Bates and John Jervis. Long before the Erie Canal was finished, Wright's reputation as a canal engineer had spread throughout the country, and his services, as a canal consultant and supervisor were sought wherever canals were being contemplated.

He was a consultant on the Connecticut River navigation, from Tidewater to Northhampton, Massachusetts; a consultant on the Delaware and Hudson Canal; a consultant, and later Chief Engineer, on the James River and Kanawha Canal; a consultant on the Blackwater Canal in Rhode Island and Massachusetts; Chief Engineer of the Chesapeake and Delaware Canal; Chief Engineer on the Chesapeake and Ohio Canal; Chief Engineer on the Delaware & Hudson Canal; Chief Engineer on the St. Lawrence Ship Canal; Chief Engineer on the Welland Canal; and a consultant on the Illinois-Michigan Canal — in the early stages of the planning and construction for each of these great waterways. Later in life, he devoted some attention to railroad engineering, in New York and Cuba.

Of Judge Benjamin Wright, Ashbel Welch, President of the American Society of Civil Engineers, said in 1882: "The skill and good judgment which was shown by this Father of American Engineering, the few errors into which he and his still more inexperienced assistants fell, the great effects produced by them with the means at their command, and the adaptation of their works to the circumstances of their time, are absolutely wonderful." In October of 1968, the ASCE declared him to be the "Father of American Civil Engineering," and in 1970 erected a bronze plaque to this effect, at his birthplace in Wethersfield, Connecticut.

JAMES GEDDES
1763 - 1838

James Geddes ("The Erie Canal" — Andrist)

James Geddes was born on a farm near Carlisle, Pennsylvania, and obtained his elementary education there. The family moved to Onondaga County, New York and settled in the town (later named in his honor) of Geddes. In 1794 he became a pioneer in the local salt industry, while at the same time studying law and surveying. He was admitted to the local bar and became a Justice of the Peace in 1800. He was elected to the New York Assembly in 1804, and was later appointed Judge of the County Court of Common Pleas. He also served in the United States Congress from 1813 to 1815.

His early surveying work, about 1808, included canal route surveys from Oneida Lake to Lake Ontario via the Oswego River; from Lewiston to the Niagara River; and from Buffalo to Seneca River. He was engaged in early construction work on the Erie Canal, running a test level between Rome and Oneida, which varied only 1-1/2" for the entire distance — remarkable accuracy for the instruments of the time. After serving briefly as the Engineer of the Western Section of the Erie, he was made Chief Engineer of the Champlain Canal, which was building simultaneously with the Erie.

Like Wright, his fame as a Canal Engineer was widespread, and upon completion of the Champlain in 1823, he was called by the State of Ohio to do a number of surveys for both the Ohio and Erie and the Miami and Erie Canals. He was the Principal Engineer on various canal surveys run along the lower Susquehanna in Maryland and Pennsylvania about 1822, and the Engineer of a canal survey in Maine, from Sebago Pond to Westbrook at about the same time, a route later to become part of the Cumberland and Oxford Canal. In 1823 he was employed by the federal government to investigate routes for the Chesapeake and Ohio Canal. He was employed on various canals being built in Pennsylvania, circa 1828.

NATHAN ROBERTS
1776 - 1852

The ancestors of Nathan Roberts were Puritans who came from England about 1640 to the Plymouth Colony. His grandfather, John Roberts, was killed in the French-Indian War, and his father, Abraham Roberts, sought his fortune (successfully) in the West Indies, but lost liberty and fortune when attempting to return to this country during the Revolu-

Nathan S. Roberts (Canal Museum, Syracuse)

tion. He was picked up by a British cruiser and forced to serve against the vessels of his own country. Abraham finally made his escape from the British, and settled in New Jersey at a place called "Piles Grove" where in 1776 Nathan S. Roberts was born.

As a young man, Nathan was able to save enough money to buy various properties in Vermont, and New York State, where he harvested timber in the summers, while teaching school at Plainfield, New Jersey in the winters. In 1804 he settled on one of his properties in Oneida County, N. Y. and taught school at Oriskany, two years later becoming Principal of the Academy at Whitesboro, same County. By the year 1816 he had become "Judge Roberts," had married the granddaughter of Judge White and had purchased a farm in Lenox, Madison County, New York, which he made his permanent home.

His activities on canals began in 1816, when, upon the solicitation of Benjamin Wright, he agreed to assist Wright in surveying the proposed route of the Erie Canal between Rome and Montezuma. The following year he was made Assistant Engineer for the same section. In 1818 he became Resident Engineer in charge of canal construction work between Rome and Syracuse. So well was his work done that in 1819 he was placed in charge of a party to locate the canal between the Seneca River and Rochester, placing the work under contract after detailed plans were complete.

He continued in charge of the construction of this section until 1822, when he was sent ahead by the Canal Commissioners to take charge of the Lockport area and superintend construction of the canal from this point to Lake Erie.

Here Roberts faced the greatest challenge of his career. The Niagara escarpment, of almost solid rock, presented a seventy-five foot barrier across the route of the canal. Beyond was a rocky plateau of seven miles length. An entire corps of engineers was already at work trying to solve the problem, when Roberts took charge. The entire success or failure of the Erie Canal (now almost completed to this point) hinged upon these last few, and most difficult miles. Without consultation with anyone, and with little aid from existing published works, Nathan Roberts drafted a plan for the proposed structure and the channel across the plateau.

In 1823 he laid his plan before the Canal Board, complete in all details of construction and operation (along with a number of other plans which his contemporaries had also submitted.) Roberts plan, though expensive, was finally selected as the most logical solution, and he was appointed by the Board to personally supervise its construction.

The famous double-five flight of locks at Lockport, New York, designed by Nathan Roberts to raise the Erie Canal to the upper level of the Niagara Escarpment.

Thus was created the famous double "flight of ten locks" at Lockport, New York, followed by a thirty-foot deep cut through nearly solid rock to tap the waters of Lake Erie. The Lockport locks were opened to traffic in the fall of 1825. Everything worked to perfection!

This triumph of canal engineering behind him, Nathan Roberts' services were in immediate demand throughout the country. First, he was called to serve on the board of consulting engineers for the Chesapeake and Delaware Canal. Next, he was employed by the State of New York (1826) to run a survey for a ship canal around Niagara Falls (a plan abandoned when the first Welland Canal was built in Canada.)

In 1827 he accepted an appointment as Chief Engineer of the western section of the Pennsylvania Main Line Canal, between Pittsburgh and the Kiskiminetas River. While engaged in this construction, and during a visit to his farm in Lenox, New York he also completed a detailed study of the summit level of the Chenango Canal for the New York State Canal Board, with particular attention to providing adequate water to the summit.

Returning to Pennsylvania, Roberts found two new assignments waiting for

Crude but effective hoisting equipment designed by Nathan Roberts and his engineer corps, to remove the rock from the Erie Canal channel west of Lockport, after blasting it loose with black powder. (The Bettmann Archive)

him — one, a review of the estimates for construction of the Chesapeake and Ohio Canal, and two, an appointment as Chief Engineer of the Pennsylvania Canal, with special attention to a Portage Railroad to connect the Juniata and Western Divisions of the system. He worked at the latter assignment until 1828. A more lucrative appointment to the Board of Engineers of the Chesapeake and Ohio Canal, sent him into Southwestern Pennsylvania to develop the location for the final connection of the C. & O. canal across Allegheny Mountains — Cumberland to Pittsburgh. The following year he extended the construction work on the C. & O. from Point of Rocks to Harpers Ferry, and in 1830 was stationed in Washington, where he supervised construction on the first division of the C. & O. Canal.

The Federal Government employed Nathan Roberts for a two-year stint (1830-32) as Chief Engineer in charge of an investigation of a ship-canal around Muscle Shoals on the Tennessee River. Returning to New York he was asked to investigate possible enlargements to the Erie Canal, and from 1835 to 1841 acted as Chief Engineer for the enlargement program, assisted by Engineers John B. Jervis and Holmes Hutchinson. During this period he rebuilt one of the tiers of the Lockport Locks and completed the Rochester Aqueduct. He retired in 1841 and died at his home in Lenox in 1852.

CANVASS WHITE
1790 - 1834

Born in Whitestown, New York, Canvass's early education included a year at Fairfield Academy, where he studied mathematics, surveying and engineering. Always in poor health, a sea voyage was recommended for him and he shipped as "super-cargo" on a merchant vessel bound for Russia. He returned in 1812, with greatly improved health, to find his country involved in a second War with Great Britain. In 1814 he organized a company of volunteers, in which he was commissioned a Lieutenant, and participated in the capture of Fort Erie. During the battle he was severely wounded.

In the spring of 1816 he was engaged by Benjamin Wright to assist in final survey work along the route of the Erie Canal. Wright and later De Witt Clinton were so impressed with the enthusiasm and natural ability of the young engineer that they suggested he make a trip to England to collect information there which would help with some of the troublesome details of the Erie. This he did, at his own expense.

He wrote his father from Liverpool in January of 1818: "I have traveled 400 miles (he later covered a full 2000), passed through a number of tunnels and over

Canvass White (ASCE Biographical Dictionary)

several aqueducts . . . One aqueduct which I have examined (the Pontcysyllte) conducts a canal across the River Dee in North Wales, and consists of 19 arches of cast iron . . . I shall now make a tour through the North of England."

He returned to the States in the Spring of 1818, with the latest surveying instruments, detailed drawings of English canal operations, and much information on the use of hydraulic cement in construction of English canal locks.

As previously mentioned, it was Canvass White who solved the problem of an inexpensive supply of hydraulic cement for the Erie Canal. It was not without considerable searching and researching on the part of the young engineer for a good American natural cement that would harden under water — that led him to a deposit of stone near Chittenango, New York. When burned and pulverized this stone produced the desired results. White obtained a patent for his new cement (valid in 1820) and a cement processing plant was set up, under the supervision of his brother, Hugh White, to produce this much needed material for the locks, aqueducts and bridge foundations for the Erie Canal, and later — many others.

As work on the Erie progressed, Canvass White became the chief expert in designing its locks and other appurtenances, which earned him the title of "Principal Engineer."

Before the Erie was complete, White was approached by a number of other canal companies to assist them in their planning and design work. He became a consultant on the Schuylkill Navigation, the Farmington Canal, the Chesapeake and Delaware Canal; and was, for a time, Chief Engineer on the Union Canal, the Delaware and Raritan Canal, and the Lehigh Canal, respectively. Ill health often interfered with his canal projects, and he finally retired to Florida, where he died at the untimely age of forty-four — a great loss to the canal builders of the early Nineteenth Century.

Running a fresh team out of the stable on an Erie Canal freight boat. (Canal Society of New York State.)

"CANAL FEVER"

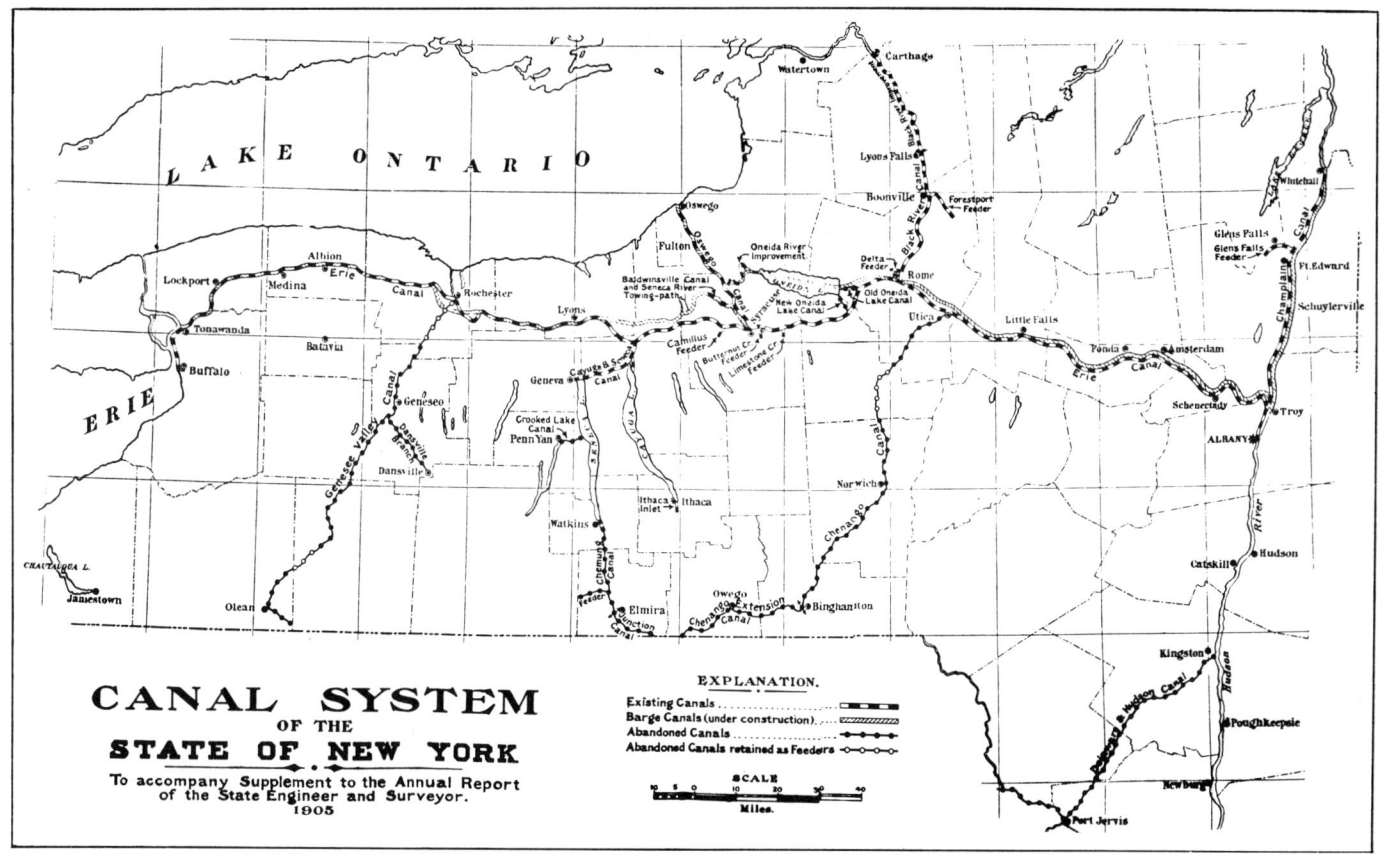

As literally thousands of boats began moving between Albany and Buffalo, it became evident that the opening of the Erie Canal was one of the most significant single events in American history since the Revolutionary War. Immigrants bound for the western states shifted their route from the arduous mountain roads of Western Pennsylvania and Virginia to the far easier, and less expensive Erie Canal. Merchandise of all kinds began moving in both directions from the "western states" of Ohio and Michigan into the East, via Buffalo, Albany and the Port of New York. Prices of eastern manufactured goods dropped dramatically in the Western States, as did prices of such luxury items as furs from the Great Lakes area in New York City. The little hamlets of Syracuse and Rochester, along the route of the Erie, became "boom towns," with Rochester producing flour for the entire Northeast.

Neighboring States watched these developments in New York with anticipation, or dismay, depending upon the beneficial or adverse influence of the Erie Canal on their own trade. In any event, the completion of the Erie Canal touched off a flurry of canal-building throughout the entire Northeast unrivaled in History since the canal-building period in the United Kingdom half a century earlier. The following is a brief account of major canal activities in other States.

MASSACHUSETTS

Bostonians, seeing their position as one of the major seaports in America slipping away to New York City, hired the younger Loammi Baldwin to run a survey for a canal straight west from Boston to the Hudson River, to tap some of the anticipated Erie Canal trade. Baldwin found one great barrier at the Berkshire Range, (later pierced by the Hoosac railroad tunnel). He figured on a four-mile canal tunnel at this point, and his estimate for the entire water route, presented to the Massachusetts Legislature in 1826, was "guesstimated" at over $6,000,000. The figure so thoroughly frightened the Legislature that work was never begun.

OHIO

Ohioans, delighted to be on the receiving end of Erie Canal trade, and with greatly increasing activities in Ohio ports on Lake Erie, also perceived that their State could provide important extensions of the Erie Canal west to Indiana, and south to the Ohio River. With this in mind, even before the Erie was completed, they hired James Geddes, whose work on the Erie and Champlain had attracted their favorable attention, to run surveys for north-south routes to connect the Ohio ports on Lake Erie, with River Ports on the Ohio River. Geddes submitted a report to the Ohio Legislature in December of 1822 showing five possible routes across the State. In 1824, Judge David Stanhope Bates, a pupil of the great Benjamin Wright, was retained to make more detailed surveys of two of the Geddes routes, which later became the Ohio and Erie Canal, (Cleveland to Portsmouth) and the Miami and Erie Canal (Toledo to Cincinnati). Bates was ultimately placed in charge of construction of all the Ohio canals, as Chief Engineer.

So anxious was the Ohio legislature to get started in the canal business that they authorized, in 1825, the start of the two canals simultaneously. Ground was broken, with much ceremony, for the Ohio and Erie, July 4, 1825 at Licking Summit, three miles from Newark, with De Witt Clinton and Governor Jeremiah Morrow in attendance. No sooner was this ceremony completed than the two Governors proceeded to Middletown, Ohio, where they repeated a similar ceremony for the Miami and Erie, July 21, 1825.

The Ohio and Erie Canal, 308 miles in length, was rushed through to completion, 1825 to 1832. During this period, other engineers who had learned their trade on the Erie were brought in, including Nathan

Roberts, William Price, Seigfried Dodge and Darius Lapham.

The Miami and Erie, started above Cincinnati, was opened in stages to the north and finally made it into the Michigan-Ohio politically contested "Toledo Triangle" in 1849, making connections with the Wabash and Erie Canal into Indiana in 1842. It was 249 miles in length.

Both canals were examples of excellent engineering, the Ohio and Erie overcoming a total rise and fall of 1206 feet, and the Miami and Erie, 890 feet. Other native engineers who learned their trade on the Ohio canals included Capt. Francis Cleveland, an uncle of Grover Cleveland, Jesse Williams, Isaac Jerome, Byron Kilborne, and Samuel Forrer.

PENNSYLVANIA

Pennsylvanians correctly perceived the New York and Ohio Canals as a flanking movement which might cut them out of western trade almost completely, and relegate the Great Port of Philadelphia — so important in American colonization of the New World — to third or fourth place. It is true they had one of the first hard-surface highways in the country — the Pennsylvania Road between Philadelphia and Pittsburgh (opened in 1820) but travel over the Allegheny Mountains was still difficult and the hauling of merchandise expensive.

A commission known as the "Pennsylvania Society for the Promotion of Internal Improvements in the Commonwealth" had formed in 1824 and had sent Engineer William Strickland to England to investigate the new railroads which were being built there. Strickland was also asked to

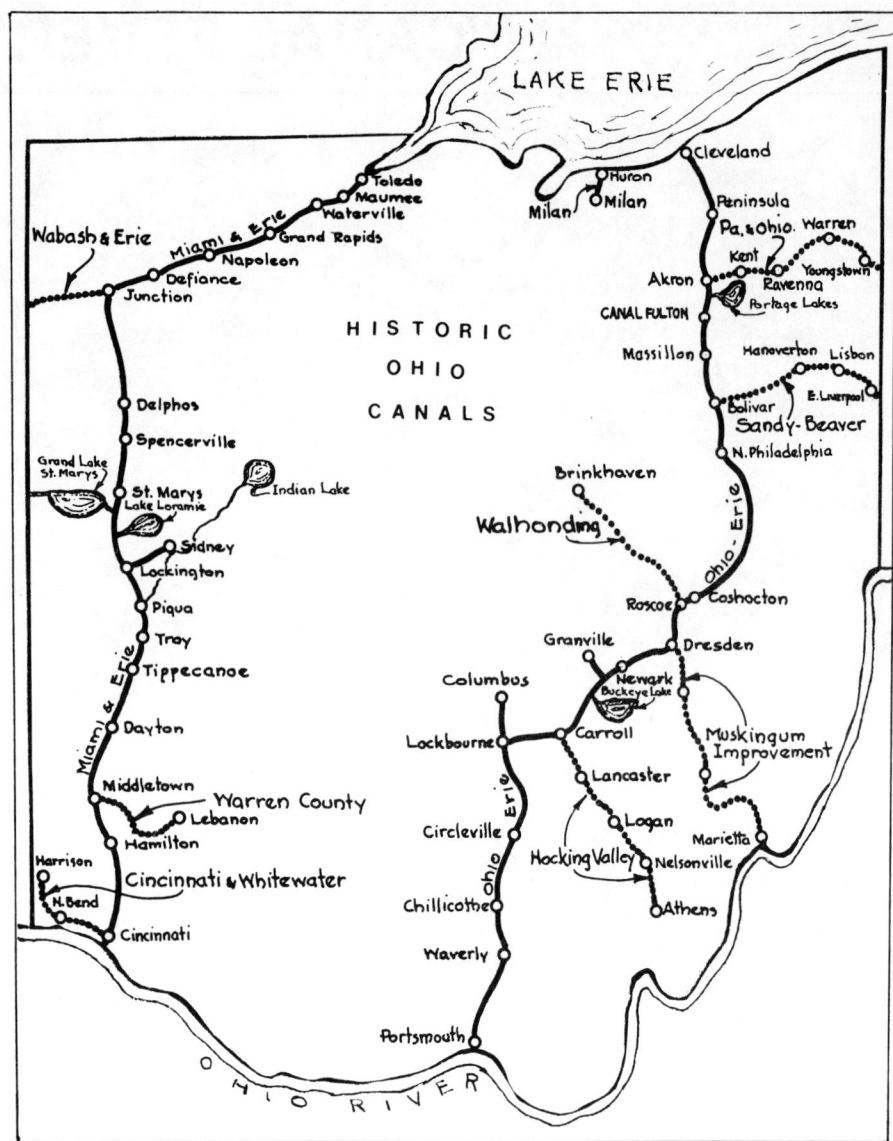

The inlet basin at the north end of the Susquehanna and Tidewater Canal, Wrightsville, Pennsylvania, Circa 1885. The S. & T. Canal formed an important connection between the canals of Pennsylvania and Maryland in the mid-1800's.

find out what progress was being made in the invention of a successful steam locomotive, to replace the horse-drawn cars then in use. Strickland returned with an excellent report, indicated that railroads would soon be the new transport media. However, the advocates of canals for Pennsylvania paid no heed to these warning signs, and pushed ahead with plans to create a canal route across the State which would rival the Erie Canal. It was felt that the Schuylkill Navigation, which had started building north from Philadelphia in 1816, could be connected with the Union Canal (Reading to Middletown) which was being rushed to completion, 1821-1828. From this point, the canal proponents envisioned a water route along the Susquehanna and Juniata, with a four-mile canal tunnel through Allegheny Mountain to connect with the Conemaugh, Kiskeminetas and Allegheny Rivers into Pittsburgh, and the Ohio River west.

The Pennsylvania State Legislators appointed a Canal Commission (1824) and in February of 1826 authorized the construction of what was later known as

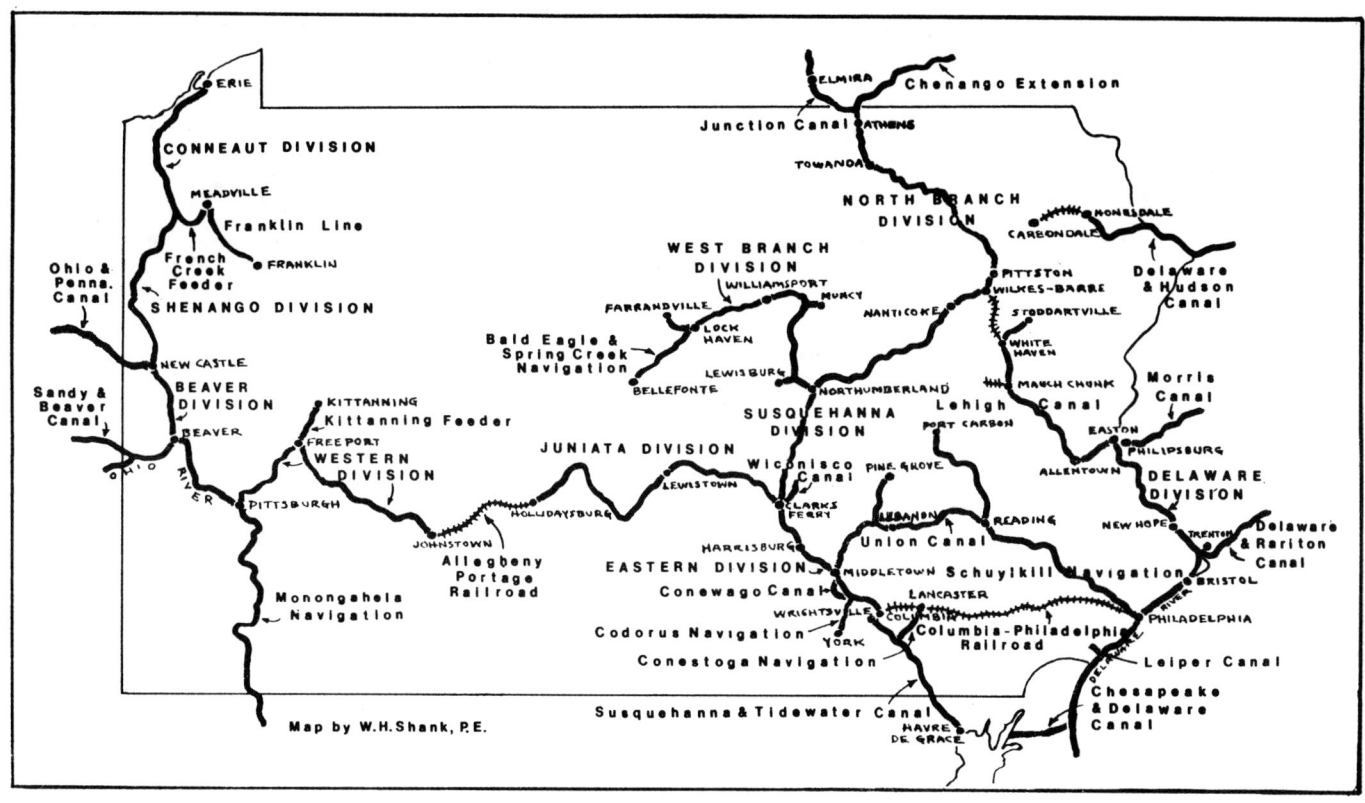

"The Main Line" between Philadelphia and Pittsburgh. Ground was broken at Harrisburg on July 4, 1826. As the work proceded, alterations were made which included a horse-powered railroad from Philadelphia to Columbia (The Union Canal was found to be much too narrow to accomodate the larger boats which soon began plying the state-owned canals) and an inclined-plane railroad over Allegheny Mountain, instead of a tunnel. The entire 395-mile route was rushed through to completion in 1834, and an extension begun to connect Beaver, on the Ohio River, with the town of Erie (Pennsylvania) on Lake Erie.

"Canal Fever" hit Pennsylvania with a vengeance and by 1830 some 1400 miles of canals, both publicly and privately owned, were planned, under construction, or completed.

Canal Engineers who had received their training on the Erie Canal were in great demand throughout the State. To try to pin-point specific engineers for particular canals is difficult, as many of them were consultants for a number of canals simultaneously, and were moving constantly from one project to another. However, the following engineers were either imported from New York, or received their training "on the job" under the supervision of Erie Engineers: Nathan S. Roberts (Pennsylvania "Main Line" Canal); William Milnor Roberts (Lehigh Canal, Union Canal, Allegheny Portage Railroad, Monongahela Navigation); Canvass White (Union Canal, Delaware and Hudson Canal, Lehigh Canal); Samuel Honeyman Kneass (Susquehanna Division, Delaware Division, Delaware and Schuylkill Canal, and Wiconisco Canal); Horatio Allen (Delaware and Hudson Canal); John Bloomfield Jervis (Delaware and Hudson Canal); and Charles Ellet (Schuylkill Navigation). There were a number of others., such as Hother Hage, Simeon Guilford, Francis Rawle, Robert Faries, etc.

A passenger-carrying canal packet boat on the Pennsylvania Main Line at Rockville, Circa 1880. The bridge in the background is the Pennsylvania Railroad crossing of the Susquehanna River northwest of Harrisburg. (Courtesy Dr. Ernest Coleman.)

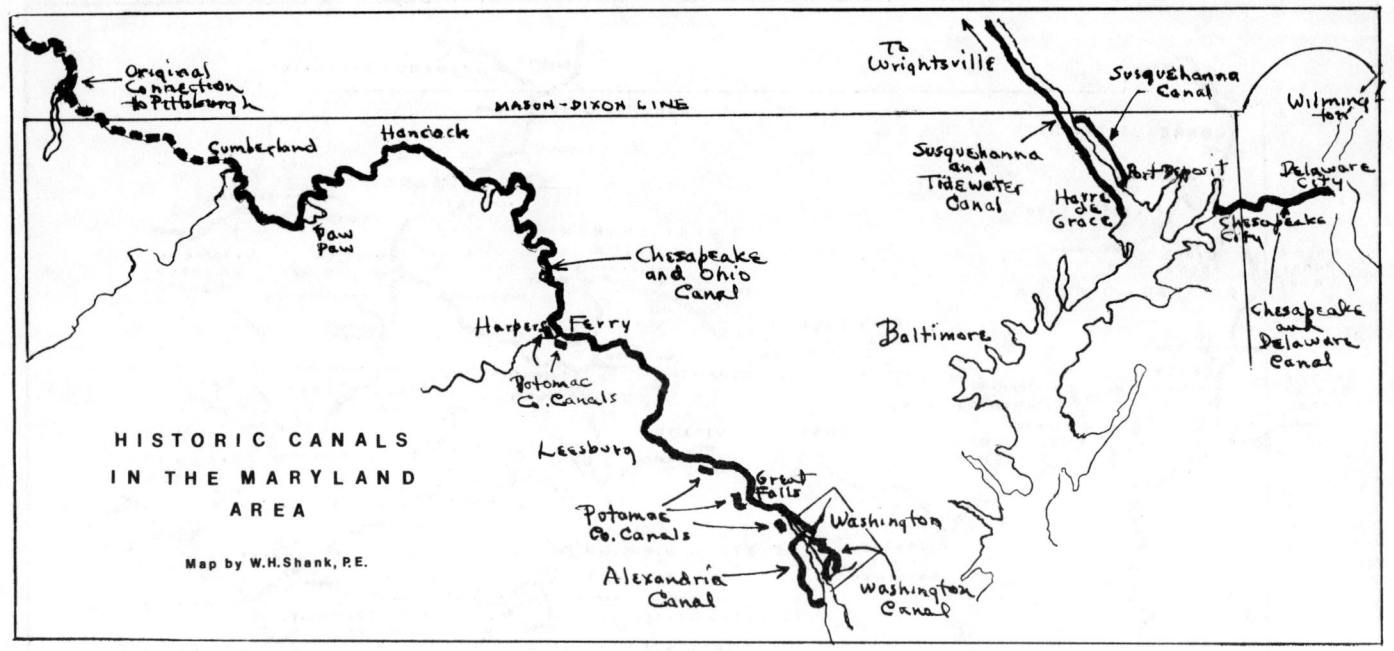

MARYLAND

As the implication of activities to the North became apparent, both in New York and Pennsylvania, Marylanders became concerned about the lack of any medium in their state to catch a share of the Western trade for the Port of Baltimore. By this time it was becoming apparent that the English were about to invent a workable steam locomotive. Railroad contests were being run there, and the Stephenson father-son team was awarded first prize for their "Rocket" in 1829 — the first fully practical locomotive in the World. So Maryland began considering both canals and railroads.

The obvious choice for a canal to connect Maryland with the West was a continuation, or a replacement, for George Washington's old Potomack Canal along the Virginia side of the Potomac River — a series of short by-pass canals around river rapids, which by 1824 had worried its way as far west as Harpers Ferry. A new company, the Chesapeake and Ohio Canal Company, was formed in 1824, and authorized to take over the works of the old Potomack Company and continue them to the Ohio River at Pittsburgh.

This company planned to build a new canal, on the Maryland side of the Potomack from Georgetown (D.C.) to Cumberland, Maryland and from this point over the Allegheny Mountains to the Monongahela River, near Pittsburgh. A four-mile, 80-yard long tunnel was planned to pierce the summit ridge in the Alleghenies. The route was slightly over 341 miles in length. Company plans also included an extension from Georgetown to Alexandria, Virginia. All plans had the blessing of both the Maryland legislature and the United States Congress.

However, before the canal company had begun work, another company was authorized to build a railroad — the Baltimore and Ohio — along virtually the same route. Galvanized into action, the new Canal Company finally broke ground in Georgetown, with the participation of President John Quincy Adams, on July 4, 1828, and construction began. On the same day, construction also began on the B. & O. Railroad in Baltimore. From this day on there was almost continual competition for right-of-way of canal versus railroad on the north bank of the Potomac. At particularly narrow points along the river bank, the work of both companies was often held up while litigation decided which company would use the prime roadbed. The canal company suffered unusual expenses and delays and finally reached Cumberland, Maryland in 1850, eight years behind the railroad, and after having spent $12,000,000 to build only 184 miles of their projected route. At this point they gave up the ghost!

The 7-mile long Alexandria connection was completed in 1843 (by a separate company) with the building of a 1000-foot long aqueduct across the Potomac at Georgetown to the Virginia side. The Washington City Canal, designed by Benjamin Henry Latrobe, and built 1810 to 1815, made connection with the C. & O. Canal via Rock Creek in Georgetown, along what was known as the "Washington Branch" of the C. & O.

The "Canal Clipper" locks through the restored upper lock at Great Falls, on the Chesapeake and Ohio Canal, with a load of passengers. Note the mule team and driver on the Towpath. (Photo by the Author.)

Another canal of importance to Baltimore in securing trade to the North, was the Susquehanna and Tidewater Canal. Baltimorians had been trying since their construction of the old Susquehanna Canal in the late 1700's to make connections into Pennsylvania, but were held up by Philadelphians, who rejected the idea of permitting inland trade with central Pennsylvania by way of Maryland. The Chesapeake and Delaware Canal changed all this, so in 1835, Pennsylvania granted a charter to the Susquehanna Canal Company, to build a joint canal with the Tidewater Company, already chartered in Maryland, and the Susquehanna and Tidewater Canal was pushed through to completion 1836-1840.

Another important canal, with part of its mileage in Maryland, was the Chesapeake and Delaware Canal, crossing the 13-1/2-mile isthmus of land between the upper Chesapeake Bay and Delaware Bay in the State of Delaware. This connection, which made Chesapeake Bay part of the Atlantic inland waterway, had been discussed for many years. Construction was finally begun in 1824 and completed in 1829, with two lift locks overcoming a summit level of 12 feet above tidewater.

Most active Canal Engineers in the Maryland area were: Benjamin Wright, who was Chief Engineer on the Chesapeake and Delaware in 1824, and Chief Engineer on the Chesapeake and Ohio in 1828-1831; James Geddes who ran surveys along the lower Susquehanna in 1823; Canvass White who worked briefly on the Chesapeake and Delaware; Charles Ellet, who was Assistant Engineer on the Chesapeake and Ohio; Samuel Kneass who was Assistant Survey Engineer on the Chesapeake and Delaware; and Edward Gill and Horatio Allen, both of whom worked on the Chesapeake and Delaware. Other engineers involved in Maryland canals included: Nathan Roberts, Walter Gwynn, William Hutton, Stephen Long, Charles Fisk and Ashbel Welch.

Typical masonry aqueduct on the lower section of the Chesapeake and Ohio Canal. (Photo by Tom Hahn.)

NEW JERSEY

The Morris Canal, across northern New Jersey, was the "highest climbing" of all the historic canals of the USA in the early 1800's. It made a connection with the Lehigh Canal at Phillipsburg, climbed 760 feet to Lake Hopatcong — the summit level, from which it drew its water supply — and then dropped 914 feet to tidewater at Newark Bay. As the crow flies, it was only 55 miles from Phillipsburg to Newark, but the Morris wound its tortuous way around the contours of the hills for over 102 miles, which put it at a disadvantage in competing with the much shorter rail lines.

To overcome a total rise and fall of 1674 feet, it would have required over two hundred locks and with the limited lock lift of the day, almost prohibitive cost problems. James Renwick, the English Engineer in charge, overcame the difficulty in a spectacular way by building twenty-three inclined planes which took care of most of the grade. Only twenty-four lift locks were needed for the rest of the route.

The planes were actually a double series of marine railways, with up-going boat-cradles balanced against down-hill cradles and extra power supplied by an ingenious underground water turbine, fed by water from the upper reach of the canal. An amazing continuous hemp rope system, later replaced by wire cable, activated the cradles.

Average lift of the planes was about 63 feet; the longest (at Boonton) 80 feet. Special boats, built for the Morris, were in two hinged sections, as were the cradles, to permit boat and cradle to pass smoothly over the "hump" of the plane and enter the upper canal level.

The Morris began construction around 1824 and was completed to Newark Bay in 1831. The first boat to pass entirely through the canal was the "Walk-in-the-Water" which arrived in Newark when the canal was opened, May 20, 1832.

A twelve-mile extension of the Morris, around the tip of the Jersey City peninsula, was completed in 1836 to bring the canal directly into the Bay of New York. The Morris continued operations until 1924.

An extremely important canal in New Jersey was (and still is) the Delaware and

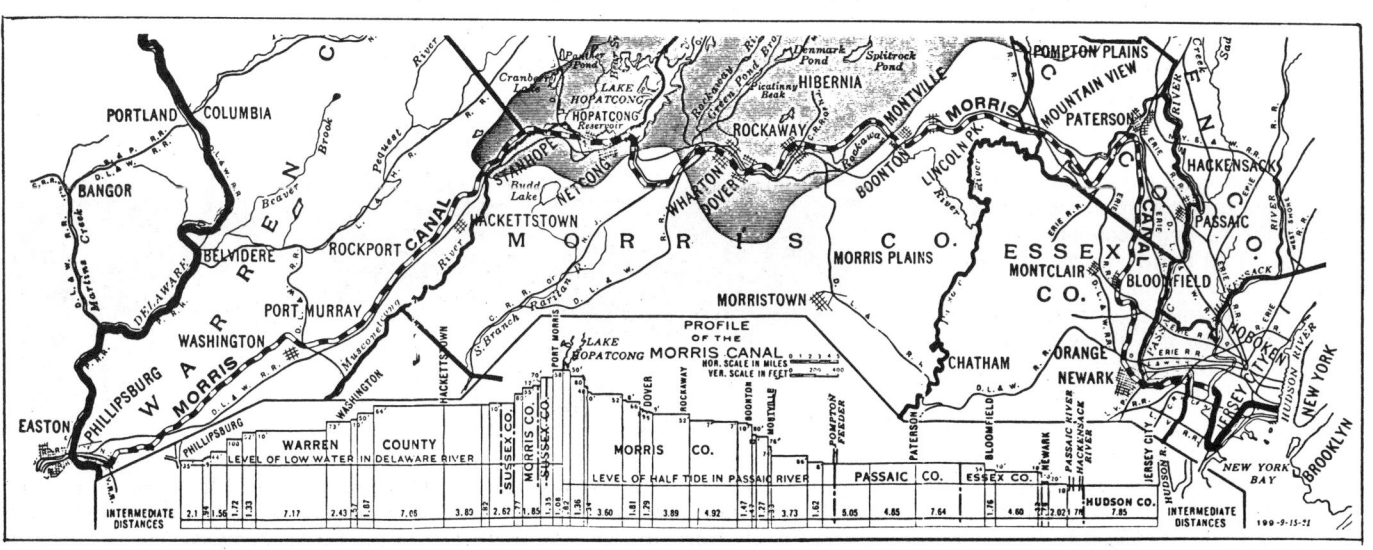

Inclined planes on the Morris Canal at Newark, New Jersey. The Morris was one of the few canals in the United States which used inclined planes instead of locks, to raise or lower the canal boats between levels.

canal was obtained from a 22-mile long feeder canal running north from Trenton to Raven Rock on the Delaware River, 50 feet wide by five feet deep, and also navigable. The main stem of the canal was 80 feet wide by 8 feet deep.

In its first ten years of operation, the major trade of the Delaware and Raritan Canal came from coal and lime boats descending the Schuylkill Navigation into Philadelphia, which were previously forced to unload there and trans-ship to New York City. Now these boats could transit the entire route by water, via the D. & R. Canal. Another large source of revenue for the new canal were the coal boats of the Lehigh Canal. Because of the limited width of the cumbersome Morris Canal, many of the Lehigh Canal boats came down the Delaware Canal from Easton, crossed the Delaware via special outlet locks to the D. & R. Feeder, to join the main canal at Trenton, completing their journey to New York via New Brunswick and the inland water route west of Staten Island.

In 1860, out of more than 1,200,000 tons of coal alone, passing the D. & R. Canal, approximately 900,000 came from the Schuylkill and 314,000 from the Lehigh. However, the Schuylkill Navigation was being given serious competition by the Philadelphia and Reading Railroad, which eventually bought them out (1870). In the meantime the Camden and Amboy Railroad had been purchased by the Pennsylvania Railroad (1871) which meant they were forced to acquire the D. & R. Canal as well. Because of the cut-throat competition of the Pennsyl-

Raritan Canal, connecting Bordentown on the Delaware River with New Brunswick on the Raritan River. If it had not been tied to the Camden and Amboy Railroad, it would have been one the most financially profitable canals in the country during the 1800's. As the main inland water link between the two great cities of Philadelphia and New York, its traffic occasionally exceeded that of the Erie Canal.

Political manipulations between the State, the promoters of the Camden and Amboy railroad, and the advocates of the Delaware and Raritan Canal, forced these competitive transportation media to consolidate into a single company in 1831. Stock was sold jointly in the two enterprises and construction of the canal and railroad began almost simultaneously — the former under the supervision of Canvass White, one of the outstanding graduates of the Erie Canal School.

As originally completed, in June of 1834, the D. & R. was 42 miles in length, with 116 feet of lockage, overcome by 14 locks, each 110 feet in length by 24 feet wide permitting it to handle unusually large vessels. Water to supply the

This photo, Circa 1880, shows a two-section canal boat stopping for repair work at the crest of an inclined plane on the Morris Canal. Both the "cradle" (which transported the canal boats from one level to another) and the boats themselves, were hinged to allow flexibility in passing over the "hump" at the top of the plane.

vania and Philadelphia and Reading Railroads, the D. & R. Canal was thereafter forbidden to pass any boats of the Schuylkill Navigation.

All of this cut down considerably on the trade of the lower New Jersey canal. Even so the D. & R. averaged more than 2,000,000 tons of freight per year between 1860 and 1880. In 1871 its net earnings were $1,202,419. It enjoyed a gradually reducing volume of business into the early 1900's. At one time there was talk of building a federally-funded ship canal parallel to the D. & R., but nothing came of it. At the present time the State of New Jersey is the owner of the nearly 100% watered D. & R. Canal, from which it secures a sizable income by selling water from the Canal to industries and small municipalities along the route.

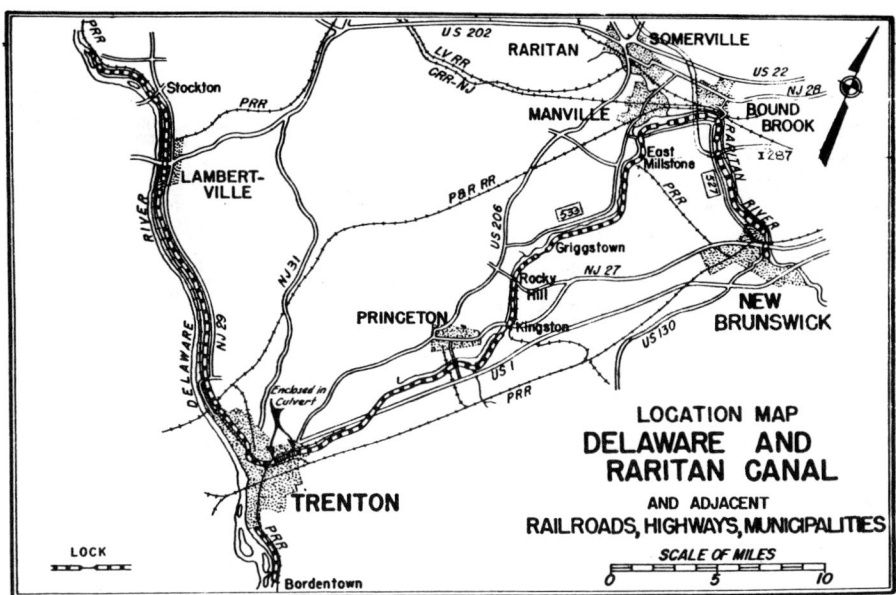

VIRGINIA

The Virginia Legislature had discussed a waterway west from Richmond as early as 1765, but it was not until George Washington made his exploratory trip into the Ohio Valley and up the Kanawha that the idea of a connecting canal across the entire State was fully promoted. Washington pushed for action, and the James River Company was formed by the State of Virginia in 1785, with Washington as honorary president. Original plans included improvement to the Falls of the James at Richmond and a turnpike road across the Alleghenies, 480 miles, to the mouth of the Kanawha River where it joins the Ohio. The seven mile canal into Richmond was completed in 1789, but after Washington's death, progress was slow, and by 1828 the canal had been built west only 28 miles to Maiden's Adventure in Goochland County plus the 7-mile Blue Ridge Canal. In 1832 the entire operation was re-incorporated as the James River and Kanawha Company, and sufficient funds had been raised by 1835 to continue construction. Judge Benjamin Wright was hired as the Chief Engineer and Charles Ellet, Jr., Simon Wright (son of Judge Wright) and Daniel Livermore were made Assistant Engineers.

Edward Hall Gill was hired in 1838 to run the surveys over the mountains and to do the detailed planning work of the route to be followed in the western section. The connecting link over the mountains to the Greenbrier, New and Kanawha Rivers had been originally planned as a highway, later as a railroad, and finally (as designed by Engineer Edward Lorraine)

Restored Lock Number Seven at Battery Creek on the James River and Kanawha Canal. In the background is the Blue Ridge Parkway Bridge. The Lock is a feature attraction in an area maintained by the National Park Service. (Photo by the Author.)

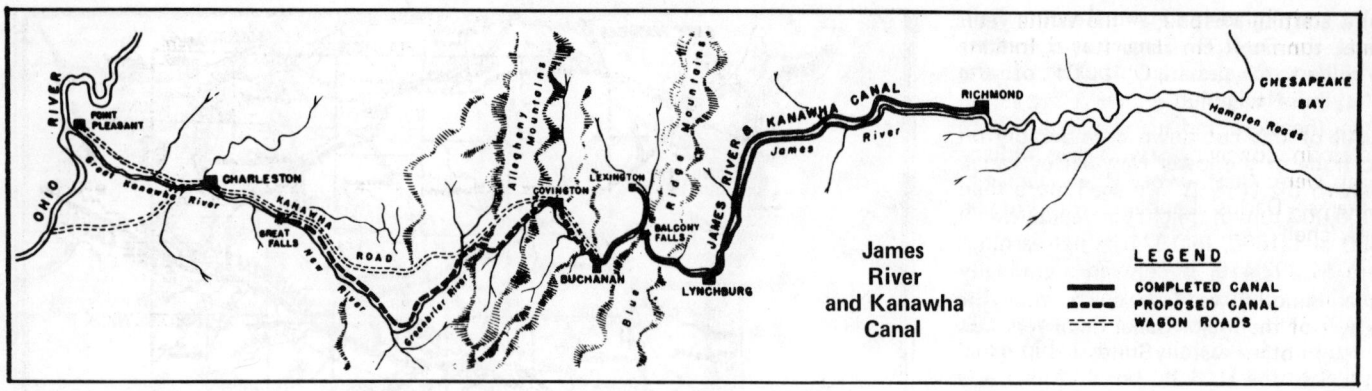

a nine-mile-long canal tunnel. Completion of the tunnel was given serious consideration by the Federal Government (1870-1874) as a "project of national importance." A number of Corps of Engineers surveys were run, but federal funds were never made available and the project finally died.

At the eastern end of the system, construction proceeded west to Buchanan, (1851) 197 miles west of Richmond, and although some expensive construction work was done further west (between Buchanan and Eagle Rock) this section was never opened to navigation. Chief Engineer in the latter days of the company operation (1847) was Walter W. Gwynn, with Edward Lorraine as Assistant Engineer. Engineer Moncure Robinson, a native of Richmond, as a young man worked on the early improvements of the James River at Richmond.

INDIANA

Caught up in the wave of canal building to the East, the State of Indiana, though thinly populated, and shortly to be served by the National Road from Maryland, nevertheless planned the longest canal yet built, anywhere in the World — the Wabash and Erie, running from a junction with the Miami and Erie Canal in northwest Ohio 397 miles to the Ohio River in southwest Indiana. Even though assisted by land grants from the federal government, the project almost bankrupted the new State!

Authorized by an act of the State Legislature in 1832, construction on the Wabash and Erie began immediately and was rushed through to completion in 1843. The State actually authorized plans for a total network of 1289 miles of canals in Indiana, many of which never got off the drawing boards. One other canal was

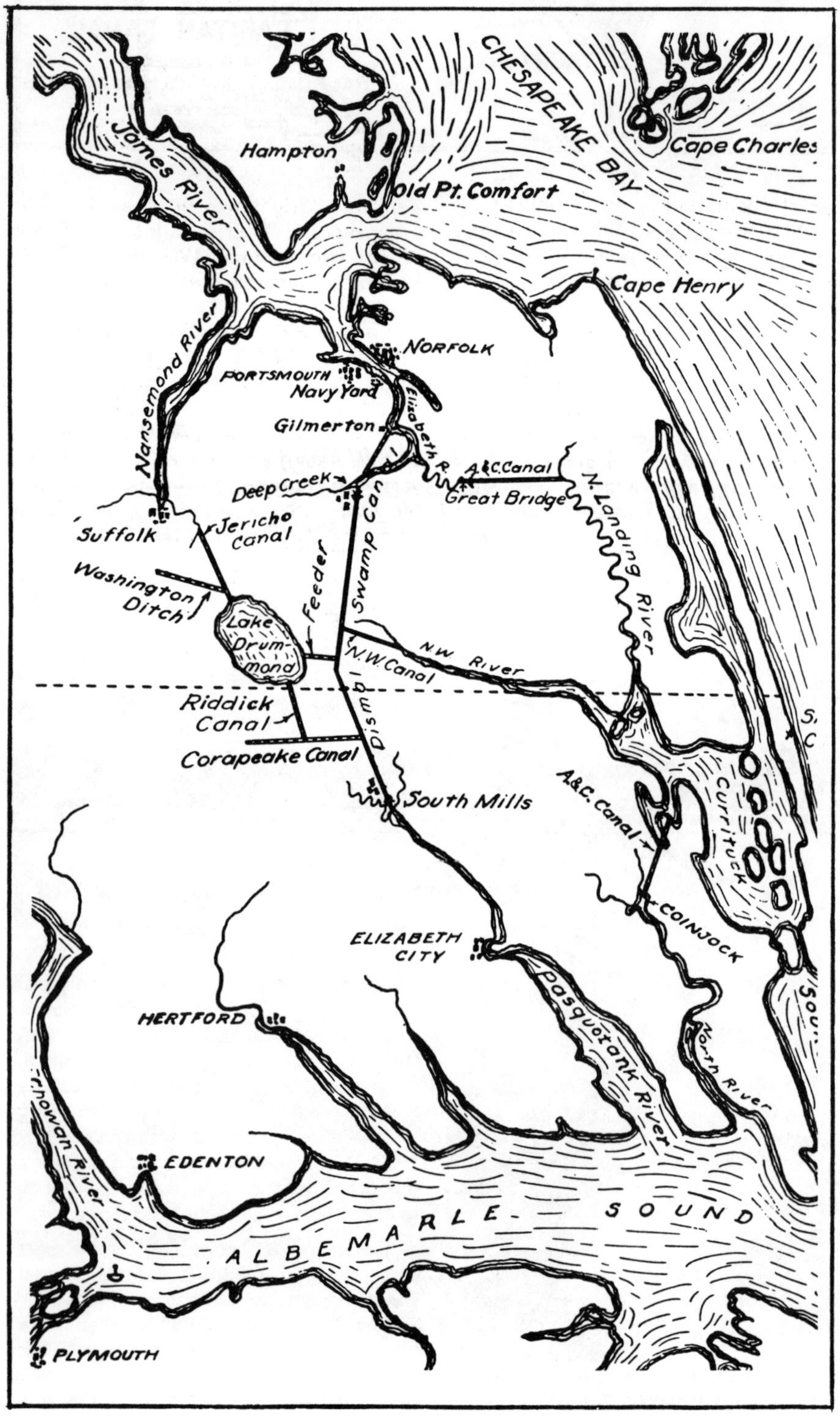

Map of the tidewater areas of Virginia and North Carolina served by the Dismal Swamp Canal and much later, by the Albemarle and Chesapeake Canal (Redrawn from an 1867 map by D.S. Dalton.)

built, starting in 1832, — the Whitewater Canal, running from Hagerstown, Indiana 74 miles to Cincinnati, Ohio. This canal is still in use today as a tourist attraction in southeastern Indiana. Engineers mentioned in connection with the Indiana canals were: Capt. James Riley, James L. Williams, Darius Lapham, and William B. Mitchell.

Before leaving the Post-Erie period of canal building we wish to examine the careers of the major Canal Engineers who served in other eastern States.

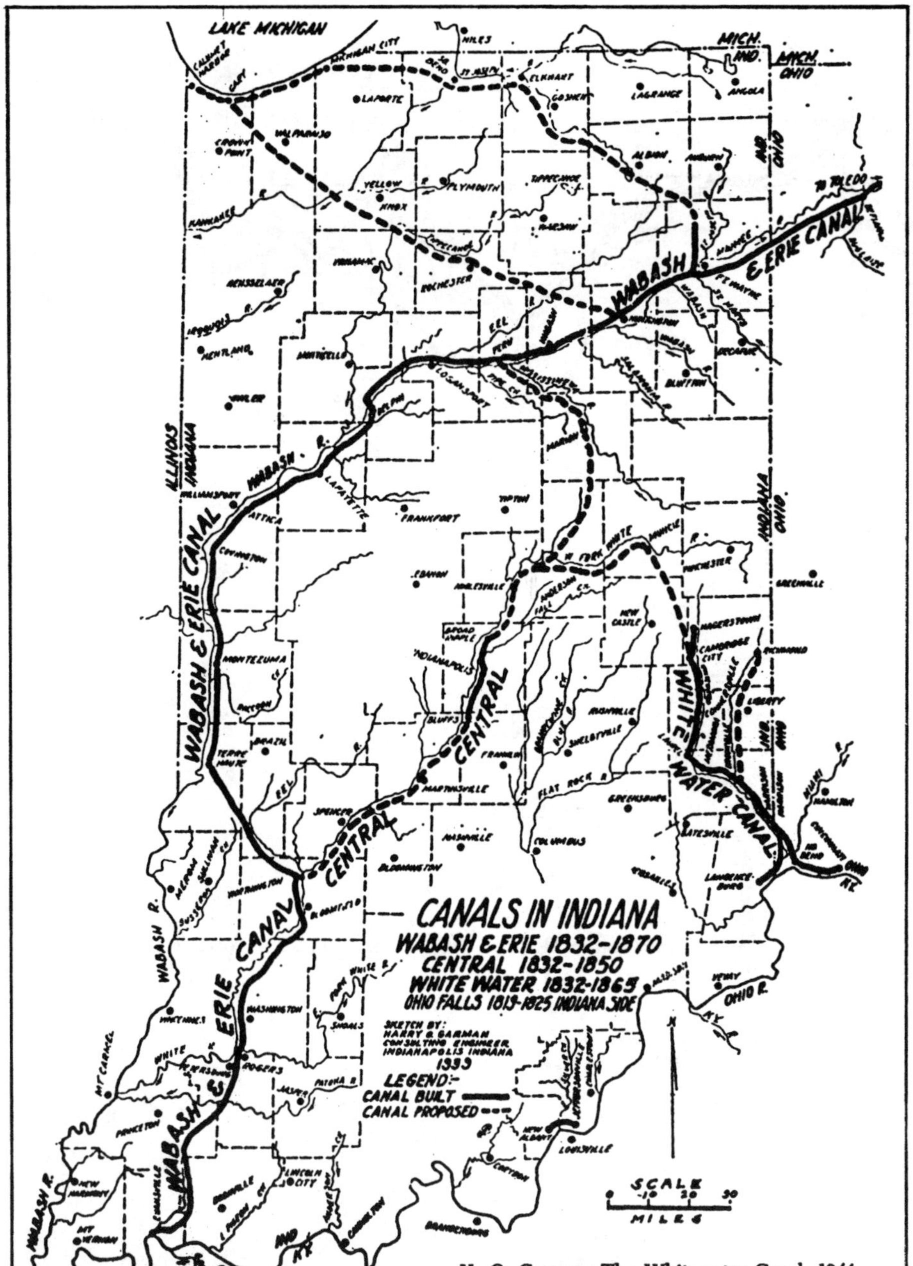

DAVID STANHOPE BATES
1777 - 1839

David Bates was born in Morristown, New Jersey, son of a Revolutionary War Officer, was trained for the ministry, but also took up the study of Mathematics and Surveying while clerking in his brother's store. Later he studied law. As a young man, he was employed by George Scriba, owner of a large tract of land in Oneida County, New York, to survey his land and act as his agent. Later, he acted as an assistant to Benjamin Wright, who was running surveys throughout Oneida County.

In the meantime, he continued his law studies, was admitted to the bar, practiced law and was elected Judge of the Court of Common Pleas of Oneida County 1811-1817. Benjamin Wright pulled him into the Erie Canal project as Assistant Engineer of the Middle Division (1817). His good work here earned him a promotion to Division Engineer in charge of the work in Ironquoit Valley, 1819-1824. He designed and constructed the 11-span masonry Aqueduct over the Genessee River at Rochester, and also worked on the construction of the flight of locks up the Niagara escarpment at Lockport, New York, under the supervision of Nathan Roberts.

Even before his work on the Erie was completed, he was retained by the State of Ohio to run surveys for the Ohio and Erie and Miami and Erie Canals there, later becoming Chief Engineer of all the Ohio Canals. His work in Ohio included 800 miles of survey work, and the opening of 200 miles of canal to actual navigation, 1824-1829. He was also Chief Engineer of the canal on the Ohio River around the falls at Louisville, 1825-1828.

Returning to New York State, Bates became Chief Engineer for the Niagara River Hydraulic Company, 1828-1834, and was also named Chief Engineer (1829) of the Chenango Canal to connect Utica and Binghamton. Later he was commissioned to survey a route for the Genessee Valley Canal.

CANALS IN INDIANA
WABASH & ERIE 1832-1870
CENTRAL 1832-1850
WHITE WATER 1832-1865
OHIO FALLS 1819-1825 INDIANA SIDE

SKETCH BY:
HARRY S. GARMAN
CONSULTING ENGINEER
INDIANAPOLIS INDIANA
1939

LEGEND:
CANAL BUILT ———
CANAL PROPOSED - - -

H. O. Garman, The Whitewater Canal, 1944

The "Valley Bell" passes through what is thought to be the only remaining covered-bridge type wooden canal aqueduct in the World, on the Whitewater Canal at Metamora, Indiana. (Photo by Eugene R. Bock.)

Delaware and Hudson Canal basin at Honesdale, about 1890. (Courtesy Richard H. Steinmetz, Sr.)

JOHN BLOOMFIELD JERVIS
1795 - 1885

John Jervis was born in Huntington, Long Island, the son of a carpenter and mill owner. His grammar school education was augmented by "on the job" experience working on the Erie Canal and continual reading of technical literature. He started as a humble axman on the Erie, was promoted to rod-man (1817) and in 1819 became a Resident Engineer in charge of 17 miles of the middle section of the Erie under the supervision of David Bates and Canvass White. When Benjamin Wright became the Chief Engineer for the Delaware and Hudson Canal (1825), he took Jervis with him as Principal Assistant Engineer. The latter became much involved with the mechanics of the gravity railroad, which brought coal from the mines at Carbondale down to the Canal at Honesdale. He assisted with the construction of the planes for the Gravity, inventing a number of devices, such as a sail device to slow the cars in their downhill travel. In this connection he also ordered the famous "Stourbridge Lion" and three other locomotives from England, used on the D. & H. Railroads. He invented one of the first American-built locomotives in 1832.

John B. Jervis
(ASCE Biographical Dictionary)

Jervis moved to the Chenango Canal, as Chief Engineer 1833-36, succeeding David Bates, designing various artificial reservoirs for the summit level of the Canal, and making surveys and estimates for the Eastern Section. He was called back to the Erie Canal as Consultant on the enlargement of the eastern division of the Canal, 1835-36. His major achievement was the construction of the Croton Aqueduct, as Chief Engineer, 1836-46, to bring water from Westchester County, on a forty-one mile continuous aqueduct (with a fall of 13-1/4" to the mile) into downtown New York. Not since the time of the Romans had anything like this been attempted. He personally supervised the construction of the high dam and reservoir on the Croton River, the many bridges and tunnels, and the construction of the aqueduct-bridge across the Harlem River. After the completion of this great work in 1842, Jervis spent the balance of his long career in railroad engineering, with time out to do some consulting work on the Boston Aqueduct (1845).

HORATIO ALLEN
1802 - 1889

Horatio Allen initially worked with John Jervis and followed him from project to project. Allen was born in Schenectady, New York. His father, being an educator himself, sent his son to Columbia College, where he was graduated in 1823, and then began a study of the law. His first experience with canals was as a rodman on a surveying party for the Chesapeake and Delaware Canal (at St. George's, Delaware) and he was later a Resident Engineer on the C. & D. (1824). He had also done some work on the Delaware and Hudson Canal, where he met John Jervis, who later made him a Resident Engineer of the D. & H. summit level in 1825. Jervis

Interior of an all-wood Aqueduct where the Pennsylvania Main Line Canal crossed the Juniata River at Amity Hall. Note the Burr-arch truss construction to support the tremendous weight of this 15-foot wide by 6-foot deep water channel.

Horatio Allen (ASCE Biographical Dictionary)

sent Allen to England to investigate steam locomotives for the D. & H. Railroads. In England, in 1828, he met George Stephenson, who, with his son Robert, was developing the "Rocket" — first successful steam locomotive in the World. Allen came home (1829) with the "Stourbridge Lion," first steam locomotive to be run in the United States. He directed construction of Jervis's own locomotive at West Point Foundry in New York City, 1829-34, and continued his interest in railroads for the balance of his career. He was also Assistant Engineer to Jervis with the work on the Croton Aqueduct.

WILLIAM MILNOR ROBERTS
1802 - 1882

William Roberts was born in Philadelphia and studied architecture at Franklin Institute. His first job was with an engineering staff under Sylvester Welch (1825) in the construction of the Union Canal. He was next a rodman in the survey work over Allegheny Mountain for a MacAdam road to connect the eastern and western sections of the Pennsylvania Canal (1826). He then moved on to the Lehigh Canal, where, as Assistant Engineer, he ran surveys and supervised construction of the canal between Mauch Chunk and Easton. In 1827-28, he also assisted Canvass White in making improvements to the inclined coal-car planes at Mauch Chunk, later known as the "Switch Back Railroad." Back to the Union Canal, he was made Resident Engineer on the Feeder Canal between Pine Grove and the "Water Works," (1830-31).

Sylvester Welch then recalled him to the Allegheny Portage Railroad, which replaced the highway he had surveyed five years earlier, and made him his Senior Assistant Engineer. With his experience on the inclined planes of the Lehigh, and some additional reconnaisance work which he was sent to do on the inclined planes of the Morris Canal in New Jersey, and the D. & H. Canal Planes at Carbondale, Pa. — Roberts became one of the leading "experts" on inclined planes in the USA. He superintended construction of eight of the ten inclined planes on the Portage Railroad (1831-34) which for the next two decades transported passengers, freight, and even entire canal boats between the Juniata Division Canal at Duncansville and the Western Division Canal at Johnstown. He was placed in charge of the entire operation of the Portage, 1834-35.

While on the Portage, William Roberts was actually on the payroll of the Pennsylvania Canal Commissioners. They pulled him out on several occasions, once to

Interior of Conemaugh Tunnel on the Western Division Pennsylvania Main Line, showing the west portal and Aqueduct, just beyond. (Hoffmann Sketch.)

Peters Creek Culvert on the James River and Kanawha Canal, just above Lock Number 8, and west of the Blue Ridge Parkway. (Photo by the Author.)

EDWARD HALL GILL
1806 - 1868

By T. Gibson Hobbs, Jr.

E. H. Gill, as he normally signed himself, was born in Wexford County, in southeast Ireland. He was the oldest son of Valentine Gill, an Irish engineer who had worked on canals in England and Ireland. The father brought his young family to this country in 1817, and served as a canal surveyor on the western section of the Erie Canal (1819).

The first record of Edward Gill as a canal engineer appears in the records of the Chesapeake and Delaware Canal. In 1825 he was an assistant engineer under Benjamin Wright, the chief engineer. In 1826 he became an assistant engineer on the Savannah, Ogeechee and Alatamaha Canal in Georgia and in 1827 was made chief engineer to succeed De Witt Clinton, Jr., who had resigned. In 1828 Gill returned north and was married later that year in New York City to Mary Schrawder of Philadelphia.

Next he served as an Engineer on the Schuylkill Navigation in Pennsylvania. Here he designed and superintended construction of parallel double locks on the Plymouth Canal section of the Schuylkill to handle the increasingly heavy traffic at this point.

In 1834, when the Sandy and Beaver Canal was organized, to connect the Ohio River in Pennsylvania with the Ohio and Erie Canal at Bolivar, Ohio, E. H. Gill was selected as Chief Engineer, with Hother Hage, a Pennsylvania Canal Engineer, as his Assistant. His first job was to re-survey the entire route (which he shortened by 18 miles) and develop plans for an adequate water supply at the summit level. The financial panic of 1837 brought work on this canal to a standstill.

In 1838 Gill came to Virginia with the James River and Kanawha Canal. The earlier James River Company, founded in 1785, with the help of George Washington, had been unsuccessful in its efforts to complete a route across the mountains to the Ohio River.

The new company was organized in 1835 to take over the original company and complete the canal, with a railroad across the mountains. Judge Wright was its first chief engineer, with Charles Ellet, Jr. as one of his principal assistant engineers. In 1836 Ellet became chief engineer and Wright stayed on as a consultant. Anxious for a good survey of the remainder of the route above Lynchburg, Ellet and Wright brought Edward Gill from Ohio in 1838 to do the work. Starting at Lynchburg, Gill completed a survey from

assist with repairs on the Western Division Canal into Pittsburgh, and later for various canal extensions on other divisions of the State-owned system, as "Chief Engineer in charge."

Now fully recognized as both a canal and railroad engineer, Roberts was never at a loss for interesting engineering assignments. He spend several years on railroad and bridge work, after turning routine operation of the Portage over to others, and then moved on to the Monongahela Navigation Company, as Engineer in Charge of construction of their slackwater dam-lock combinations, 1838-40.

He became an international figure when, in 1841-42, he was called to Canada to assist on the Welland Canal improvements. In 1843-44 he went back to northwestern Pennsylvania to help complete the Erie Extension Canal between Beaver and Erie. He then moved west to become Chief Engineer and Trustee's Agent for the Sandy and Beaver Canal in Ohio, 1845-48.

With the decline of canal-building activities, he went back to building railroads, and later became involved with such major civil engineering projects as improvements to the Ohio River (1866-70); navigational improvements to the mouth of the Mississippi River; a study of the Pittsburgh Waterworks and a similar study of the Philadelphia Waterworks. He died in Brazil where his final days were spent examining ports and waterworks for various cities there.

Chockoyotte Aqueduct on the Roanoke Canal outskirts of Weldon, N. C. (Sketch by Michael Southern, Historical American Engineering Record.)

there to the headwaters of the James. He then crossed over the mountains with his party and surveyed the New and Kanawha Rivers, on to Point Pleasant, where the latter joined the Ohio River. Judge Wright joined him on the latter part of this survey to review the work. At the same time a second party was surveying for the railroad route over the mountains.

So pleased was the company with Gill's work that he was made Principal Assistant Engineer (1839) in charge of construction above Lynchburg. That same spring Ellet lost his job as chief engineer and was replaced by Judge Wright, then nearly 69 years old. In 1840 Wright, in poor health, took a three month leave of absence, and Gill was made acting Chief Engineer. This gave Gill full responsibility for completing the 146 mile line from Richmond to Lynchburg, in addition to the newer work above Lynchburg. The first division was completed in December 1840 soon after Wright returned from New York. Gill in 1841 also made a survey of the Rivanna River north to Charlottesville which Wright reviewed and suggested the possibility of using steam boats.

By 1842 company finances were strained. Wright retired and died at his home in New York in August. Gill took over, closed out the contracts, and shut the new work down by the end of the year. He was made superintendent of maintenance and company engineer and continued to live in Lynchburg. In 1846 Walter Gwynn was elected president to succeed Joseph C. Cabell. In 1847 Gwynn was made chief engineer also, after persuading the state to fund the work above Lynchburg. He appointed Gill principal assistant engineer to direct the work. Gill refused the position, probably because he felt he should have been made chief engineer. However, in 1849, Gill changed his mind and accepted an engineering position. He was placed in charge of the Tidewater Connection work at Richmond to connect the basin with the ship dock and river below the falls.

In 1852 E. H. Gill left canal work, after 27 years, to become superintendent of the Virginia Central Railroad. He served in this capacity with three Virginia Railroads in both Richmond and Lynchburg until his death in Richmond in 1868.

CHARLES ELLET, JR. 1810 - 1862

Born in Penn's Manor, Pennsylvania, Charles Ellet early in life became interested in a career in engineering. At the age of 17, he was employed as "Assistant Engineer" with a survey crew which explored the route for the North Branch of the Pennsylvania Canal, 170 miles, from Northumberland to the New York State line. Following this he was again "Assistant Engineer," on the Chesapeake and Ohio Canal, which was starting to build west from Georgetown. He applied in 1829 for an engineering job with the Illinois and Michigan Canal, but his application was ignored, when it was learned he was under 20 years of age. Even though young Ellet's unusual talents in the engineering field were apparent, Ellet decided

Charles Ellet, Jr., about 1850.

that training abroad was essential to his career. He went to France in 1830, where he studied engineering at the "Ecole des Ponts et Chaussees." While in France, Ellet visited the Canal du Midi, with particular attention to the dam and reservoir at St. Feriol. He also studied suspension bridges being developed in France.

Ellet's 1842 wire-suspension bridge at Philadelphia, across the Schuylkill. The double outlet locks of the Schuylkill Navigation appear in the foreground. (Courtesy of Dr. Emory L. Kemp, West Virginia University.)

Returning to the United States, he was offered his old job on the Chesapeake and Ohio Canal at a much increased salary — which he turned down. By this time he was interested in suspension bridges and made an unsuccessful bid for the building of one across the Potomac at Washington. He also did some railroad survey work for Benjamin Wright in upper New York State.

In 1835, he followed Benjamin Wright to the newly re-organized James River and Kanawha Canal Company. When Wright was appointed Chief Engineer of that enterprise, Ellet went in as one of his assistant engineers, ultimately doing much of the detailed survey and design work along the lower section of the Canal. The Directors, recognizing Ellet's technical ability, sent him on several information-gathering tours to the Chesapeake and Ohio Canal and the "Main Line Canal" in Pennsylvania. When Wright resigned his position as Chief Engineer of the J. R. & K. in 1836 due to "advancing age and bodily infirmity" Charles Ellet was elevated to Chief Engineer, at the tender age of 26.

For the next three years, Ellet was the undisputed "boss" of both the design and construction of the J. R. & K. While the work was brilliantly done, he earned the animosity of both subordinates and superiors with his unwillingness to discuss and compromise with ideas not his own. Giving him full credit for the technical soundness of the work he had done for them, the directors nevertheless fired Ellet in 1839, and brought the aging Judge Wright back to replace him.

Ellet now turned his attention again to bridges, and was successful in constructing the first wire suspension bridge in the USA over the Schuylkill in Philadelphia, 1841-42 — a major engineering accomplishment!

Ellet re-entered the canal field when he became interested in the reconstruction work planned for the Schuylkill Navigation Company in 1841. The company was by this time engaged in a battle with its prime competitor along the Schuylkill (the Reading Railroad) and was trying to obtain much needed funds for necessary enlargements of its facilities to meet and hold greater traffic by water. Sympathetic to the problem, Ellet "signed on" as their agent in 1845, conducted a publicity campaign for them and a highly successful fund-raising program. As a reward for his services he was elected President of the Company. He also served as Chief Engineer on the ensuing improvements (1845-48).

Except for a brief period in 1858 when he was re-hired by J. R. & K. to plan reservoirs for navigation of the Kanawha River, the Schuylkill was Ellet's last major canal project. After this he devoted his attention to bridges and railroads. During the Civil War he invented a battering ram for steamships, and was in personal charge of several of these rams during the battle of Memphis. He was credited with a large portion of the Union naval victory there. Unfortunately, a bullet hit him in the knee and he died several days later of his wound, at the age of 52.

PRINCIPAL HISTORIC CANALS OF THE EARLY 1800'S AND THEIR ENGINEERS

Erie Canal
Benjamin Wright
James Geddes
Nathan S. Roberts
Canvass White
John B. Jervis
Charles Broadhead
David S. Bates
Theodore D. Judah
Valentine Gill
David Thomas
William J. McAlpine
Squire Whipple
William Peacock

Champlain Canal
James Geddes
Marc Isambard Brunel

Delaware & Hudson Canal
Benjamin Wright
John B. Jervis
Canvass White
Horatio Allen
John Roebling
James Renwick
Russell F. Lord
Col. John L. Sullivan
James Archibald
James S. McEntee
Portus R. Root
John T. Clark

Ohio State Canals
James Geddes
David S. Bates
Nathan S. Roberts
Loammi Baldwin II
Darius Lapham
Jesse L. Williams
Timothy G. Bates
Capt. Francis Cleveland
Isaac Jerome
Byron Kilbourne
Samuel Forrer
William Gooding

Schuylkill Navigation
Canvass White
Edward H. Gill
Charles Ellet, Jr.
Lewis Wernwag
Samuel H. Kneass
Ephraim Beach

Lehigh Canal
Josiah White
Canvass White
William M. Roberts
Sylvester Welch
Solomon W. Roberts
Alfred P. Boller

Union Canal
Canvass White
Loammi Baldwin II
William Milnor Roberts

Pennsylvania State Canals
Nathan S. Roberts
Canvass White
John Roebling
Sylvester Welch
Moncure Robinson
James Geddes
William Milnor Roberts
Solomon W. Roberts
Hother Hage
William H. Wilson
Samuel H. Kneass
Simeon Guilford
Francis W. Rawle
A.B. Waterford
Alonzo Livermore
Stephen H. Long
William R. Maffet
J. Bennett Smith
Thomas S. McNair
Robert Faries

Morris Canal
James Renwick
Asa Whitney

Delaware & Raritan Canal
Canvass White
William Strickland
Ashbel Welch

Chesapeake and Ohio Canal
Benjamin Wright
James Geddes
Charles Ellet, Jr.
Nathan S. Roberts
Horatio Allen
William Rich Hutton
Gen'l Simon Bernard
Jonathan Knight
Charles B. Fisk
Elwood Morris
Col. John J. Abert
Col. James Kearny
Maj. Walter Gwynn

Savannah & Ogeechee Canal
Benjamin Wright
DeWitt Clinton, Jr.
Edward H. Gill
Col. Alfred Cruger

Sandy & Beaver Canal
William Milnor Roberts
Edward H. Gill
John Roebling
David Bates Douglas
Maj. D. B. Roberts
Hother Hage
Joshua Malin
Washington Gill

Chesapeake and Delaware Canal
Benjamin Wright
Canvass White
William Strickland
Horatio Allen
Edward H. Gill
Ashbel Welch
Stephen H. Long
John Randel, Jr.
Samuel H. Kneass

James River & Kanawha Canal
Benjamin Wright
Charles Ellet, Jr.
Edward H. Gill
Loammi Baldwin II
William G. McNeill
Edward Lorraine
James M. Harris
Washington Gill
Maj. Walter Gwynn
Claudius Crozet
William Lake

Indiana State Canals
Capt. James Riley
James L. Williams
William S. Mitchell
Darius Lapham
William Gooding

Chenango Canal
John B. Jervis
Nathan S. Roberts
David S. Bates
William J. McAlpine

Genessee Valley Canal
David S. Bates
Frederick C. Mills

CANAL ENGINEERING

To this point we have discussed the history of canals, from antiquity to the early nineteenth century, and the lives of the men who built the canals — but little about the actual details of surveying, planning, construction and operation of the early canals.

Unlike rivers, which have a gradually descending gradient from source to mouth, canals consist of a series of "levels," connected by locks which provide for a change of elevation when necessary. Since water always seeks its own level in filling a trench or pool, each "level," whether it be a mile long or ten miles long, is exactly that — completely **level** from one end to the other. Thus, the elevation above sea level of the water level in the outlet channel of one lock is exactly the same as the inlet water level of the next lock even if the locks are twenty miles apart. This principle is well known to all Canal Engineers and enables them to run an accurate "level line" for the canal they are surveying, and make a decision as to where to locate the locks, as the topography of the country through which the canal is to pass climbs a hill or drops into a valley. In general, Canal Engineers try to pick a route which will require the minimum amount of locking from one level to another, always keeping in mind the need for a local source of water to keep each level sufficiently full to permit loaded canal boats to pass through without "dragging bottom." Many of the historic canals in America followed along a river or sizable stream, both because of the relatively easy natural gradient, and also for a ready source of water.

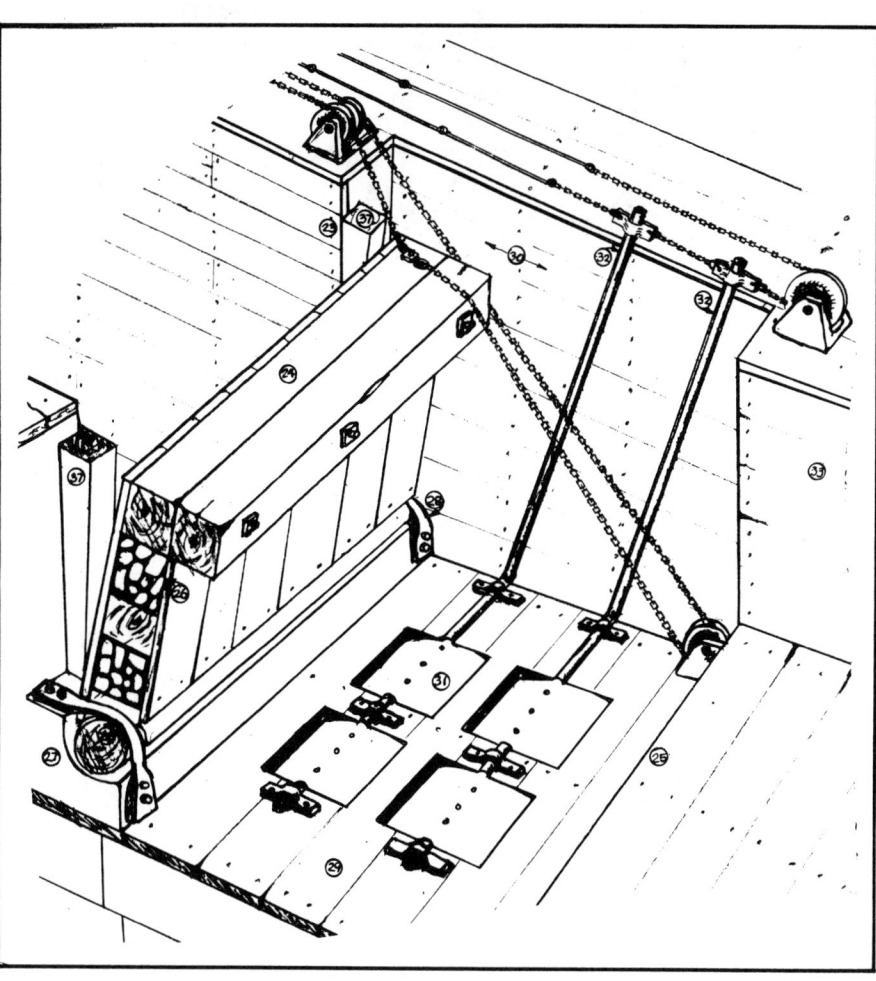

Details of operations of a "drop-type" lock gate as used on the upper end of a number of canal locks in the northeast. Sketch by Edwin D. LeRoy, from his book "The Delaware and Hudson Canal - A History, 1980."

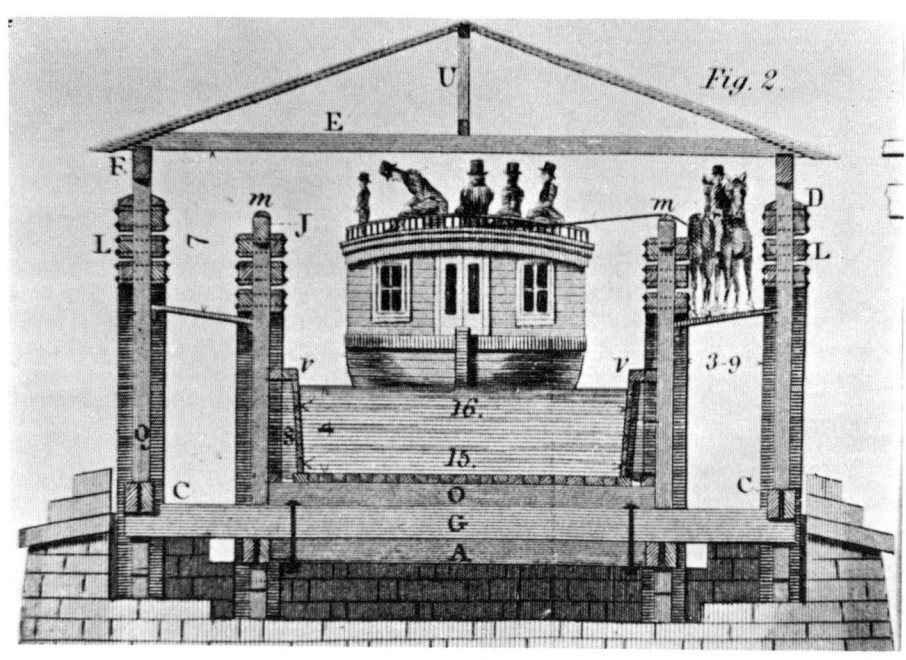

Patent drawing of the wooden Pittsburgh Aqueduct on the Pennsylvania Main Line. Originally developed by Lothrop in 1829, it was replaced in 1845 by John Roebling's first cable-suspension structure. (Courtesy Dr. Ernest Coleman.)

Thus the need for accurate "leveling" instruments in surveying the route for a canal becomes readily apparent. The best instruments were originally developed in England, and were imported into this country in the early 1800's. Any surveyor who expected to run lines for a canal, had to develop the highest possible accuracy in measuring not only horizontal distances, but in maintaining an accurate "level" line, sometimes over hundreds of miles of proposed canal route, and often through forests and hilly terrain which required constant telescopic "foresights and backsights" and great care in setting up his leveling instrument at each new location.

Then came the "profile maps" of the proposed route, and the deicision as to the location of the locks which would require the minimum amount of ground excavation for each level between them. Samples of the sub-soil conditions in the proposed canal route were often necessary. It was not feasible to run a canal through rocky soil, if a soft-soil condition existed along a slightly different route. For-

This giant device was developed by Erie Canal Engineers to remove stumps from the canal channel. A team of horses hauled on a rope coiled on the central wheel; this turned the axle, tightened the chain, and uprooted the stump. The large side wheels were used to move the device to the next stump. ("The Erie Canal" — Ralph Andrist.)

An Irish construction team digging a channel for one of the Canals.

This patented pivot-bridge was shoved aside by a moving canal boat, but returned to normal position automatically after the boat passed through.

tunately most of the river "bottom land" through which many of the canals passed was easily excavated.

Next came the estimate of building costs, which it seemed was almost always too low, as unforeseen difficulties with manpower, or building supplies, developed along the proposed route. The best locks were built of cut-stone masonry, using water proof cement. To save money some locks were built of rough stone and lined with wood. Another problem was leakage of water out of the canal channel. This was overcome by lining the canal channel with a mixture of water and clay, known as "puddling." Dams had to be planned (either full-river crossings, or "wing dams") to feed water into the canal at various points. Sometimes "slack water navigation" was found expeditious by running the canal channel in the river itself, between dams.

All this pre-planning of a canal route and the various structures to operate it, had to be carefully worked out on paper, and cost estimates submitted to the Commission or other authority responsible for financing each canal project, before construction could begin. This was the Engineer's responsibility. In most cases it also became his responsibility to hire all the contractors, stone masons, carpenters, to excavate the channel and build the locks, not to mention the bridge-builders who had to provide a canal crossing at various points where the route cut across an existing highway, or cut a farmer's grazing field in two, or where the canal itself crossed another stream. These latter structures were called "Aqueducts."

Condemnation or easement proceedings were usually the Engineer's responsibility in satisfying the land-owners along the canal route. All of this added to the cost of the canal.

The accompanying sketches and illustrations show some of the innovations in construction equipment developed by the early American Canal Engineers, as well as some of the conventional devices necessary to the operation of any canal, such as waste-weirs, aqueducts, culverts, lock gates, bridges, dams, tunnels, etc. In some cases, inclined planes were substituted for canal locks to raise or lower the canal boats from one level to another. The Morris Canal in New Jersey used inclined planes, rather than locks, over a good portion of its route.

Finally, of course, it was the Canal Engineer who was responsible for putting all these devices together, supervising the work of construction, and ironing out the inevitable "bugs," to create the final product — a working canal, which would permit canal boats, loaded with passengers or freight, or both, to pass through the hinterland of northeastern United States, where no such transport media previously existed!

HYDRAULIC CANAL CEMENT

Hyrdraulic cement may be divided into three classes, according to the method of manufacture: Portland Cement, Natural Cement and Pozzuolana. The first two must be roasted before they acquire the property of hardening under water, while the third needs only to be pulverized and mixed with water.

Portland Cement is made by high-temperature calcining an 80-20 mixture of carbonate of lime, and clay, which must be reduced to a powder by grinding, both before and after burning. It derives its name from the resemblance of the hardened mortar (made from it) to a stone found in the Isle of Portland, off the south coast of England.

Natural Cement is produced by low-temperature calcining of either a natural argillaceous limestone or a natural magnesium limestone, without pre-pulverization or the addition of other materials. After burning, the cement is then crushed into small fragments and pulverized. In Europe agrillaceous limestone is generally used, and the product is called Roman Cement. In the United States, magnesian limestone is usually employed — also called Rosendale Cement, from the place where it was first made in this country — Rosendale, Ulster County, New York.

Pozzuolana is a term applied to a combination of silica and alumina, which when mixed with common lime and made into mortar, has the property of hardening under water. The original material was discovered near the base of Mount Vesuvius at Possuoli, Italy — hence the name. It was extensively used by the Romans. Trass is a volcanic earth closely resembling Pozzuolana, and is employed in substantially the same way. It is found on the Rhine between Mayence and Cologne and in various locations in Holland.

A masonry aqueduct on the Pennsylvania Main Line Canal at Newport.

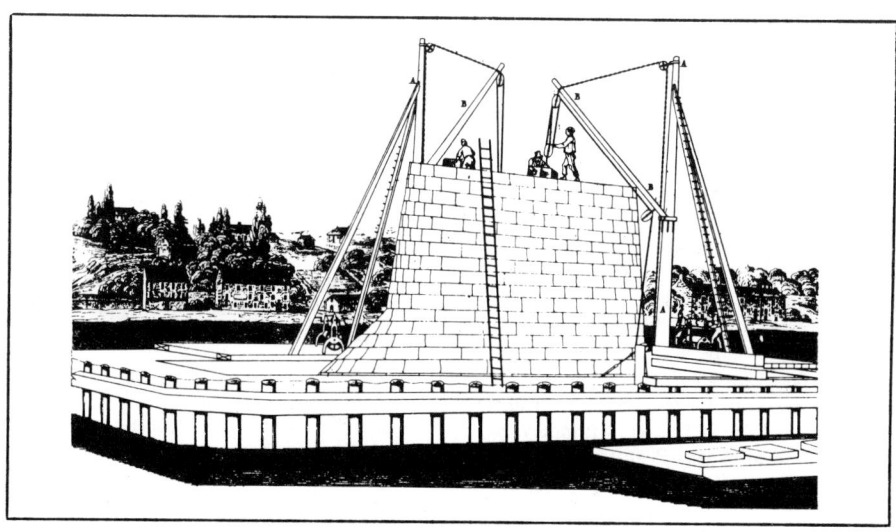

Cofferdam and pier construction for the Georgetown Aqueduct in the Potomac River, part of the Alexandria Canal project. (Courtesy "Echoes of History")

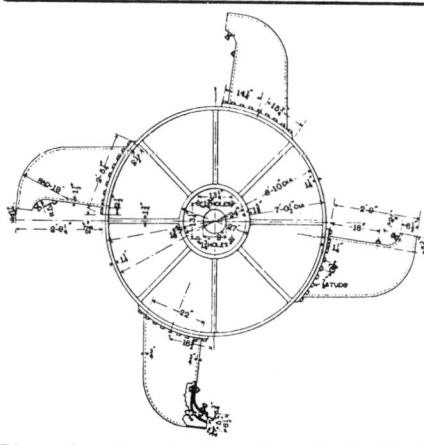

Plan of water-wheel used at the inclined planes on the Morris Canal.

The Union Canal Tunnel, west of Lebanon, Pennsylvania — oldest existing tunnel in the United States — now preserved by the Lebanon County Historical Society. (Photo by the Author.)

COAL-CARRYING CANALS

Many of the tow-path canals we have described, to this point, were intended to provide an avenue for passengers and goods between the Atlantic coastal states and the new "western states." However, there were a number of canals in Pennsylvania, New York, New Jersey, Maryland and Virginia which were originally planned (or later became) waterways for the transport of coal. Most of these canals survived the competition of the railroads for a half century or more.

One of the earliest of these was the Schuylkill Navigation, built 1815-1825 to bring anthracite coal from east-central Pennsylvania, starting at Port Carbon, into Philadelphia. Ownership of this canal was acquired in 1870 by the Philadelphia & Reading RR Company. Another was the Lehigh Canal, built 1818 to 1829, loading coal from the famous "Switch Back" Railroad at Mauch Chunk, and bringing it down to Easton, where it made connection with the Pennsylvania State-owned Delaware Division Canal, built 1827-1832, to carry coal into Philadelphia. After the State sold its interest in the Delaware Divsion, the Lehigh and Delaware canals were combined and operated by the Lehigh Coal and Navigation Company in Bethlehem. The Morris Canal, using inclined planes instead of locks at many points, was built across northern New Jersey, 1825-1831, to connect with the Lehigh Canal at Phillipsburg. The Morris ran along the contours of the hills into Newark, and was later (1836) extended into Jersey City to deliver its coal boats directly across the Hudson to Manhattan Island.

Coal boats on the Lehigh Canal, in this old print, are being checked out by the weighmaster, close to the loading point at Mauch-Chunk. The famous "Switch-Back Railroad" of the Lehigh Coal and Navigation Company was used in bringing anthracite coal down from the mountains, loading into the company coal boats on the canal.

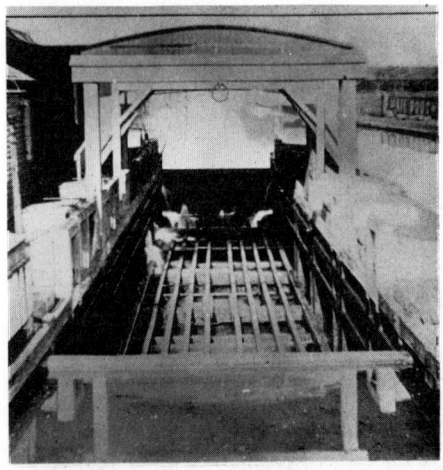

A weigh-lock at Harrisburg, Pa., showing the platform on which a canal boat rested while being weighed. A linkage to the platform showed the total weight of the boat and its coal. Toll was charged on the basis of cargo weight — computed by subtracting the "empty weight" of the boat from the weigh-lock reading.

Another coal-carrying canal was the Delaware and Hudson, built 1823 to 1828, using an amazing system of gravity railroads to bring coal from the rich anthracite deposits in northeastern Pennsylvania to Honesdale and Hawley and thence by canal boat to Kingston on the Hudson River. These canal boats were then towed "in clusters" down the Hudson to New York City.

The North Branch Susquehanna Canal, finished 1828 to 1831, to the important anthracite coal distribution center at Nanticoke, Pennsylvania, carried coal boats down the Susquehanna and Eastern Divisions of the Pennsylvania canal system to Harrisburg and made connections at Columbia-Wrightsville with the Susquehanna and Tidewater to Baltimore. Up to this time all iron had been smelted with charcoal and the nearby forests were being depleted. In 1840 the ironmasters learned to smelt iron with anthracite coal. The whole iron industry along the Susquehanna boomed, especially the furnaces at Steelton, Columbia and Safe Harbor. The bituminous coal regions of Western Maryland and Virginia (later West Virginia) were tapped by the Chesapeake and Ohio Canal, with short railroads bringing coal to the western terminus of the canal at Cumberland. Coal fields further west provided soft coal to the great steel mills and the metropolis of Pittsburgh, via the Monongahela Navigation. The latter was organized in 1817, with construction well underway in 1842, and was eventually extended into Fairmont, (West) Virginia.

It is interesting to note that almost all of these coal canals survived well into the Twentieth Century. The reason was simple: coal boats often traveled in tandem, with a single team of two or three mules on the towpath pulling with ease a load of nearly 300 hundred tons of coal. A similar load of coal by rail would have required a much greater number of cars, and an expensive share of the locomotive steam power. The truth of the matter was — it was much cheaper (per ton) to transport coal by canal boat than by railroad car. Realizing this, some of the railroad companies acquired canal lines and kept them in operation as long as canal maintenance did not become a problem.

The Schuylkill Navigation did not shut down commercial operations until 1931; the Lehigh Canal, not until 1942. The Chesapeake and Ohio Canal continued to deliver coal boats to Georgetown until 1923. The Susquehanna Canal system, under the ownership of the Pennsylvania Railroad, delivered coal to the transfer point at Columbia as late as 1901. The Monongahela Navigation, under the Corps of Engineers, still brings West Virginia coal into Pittsburgh.

CANADIAN CANALS

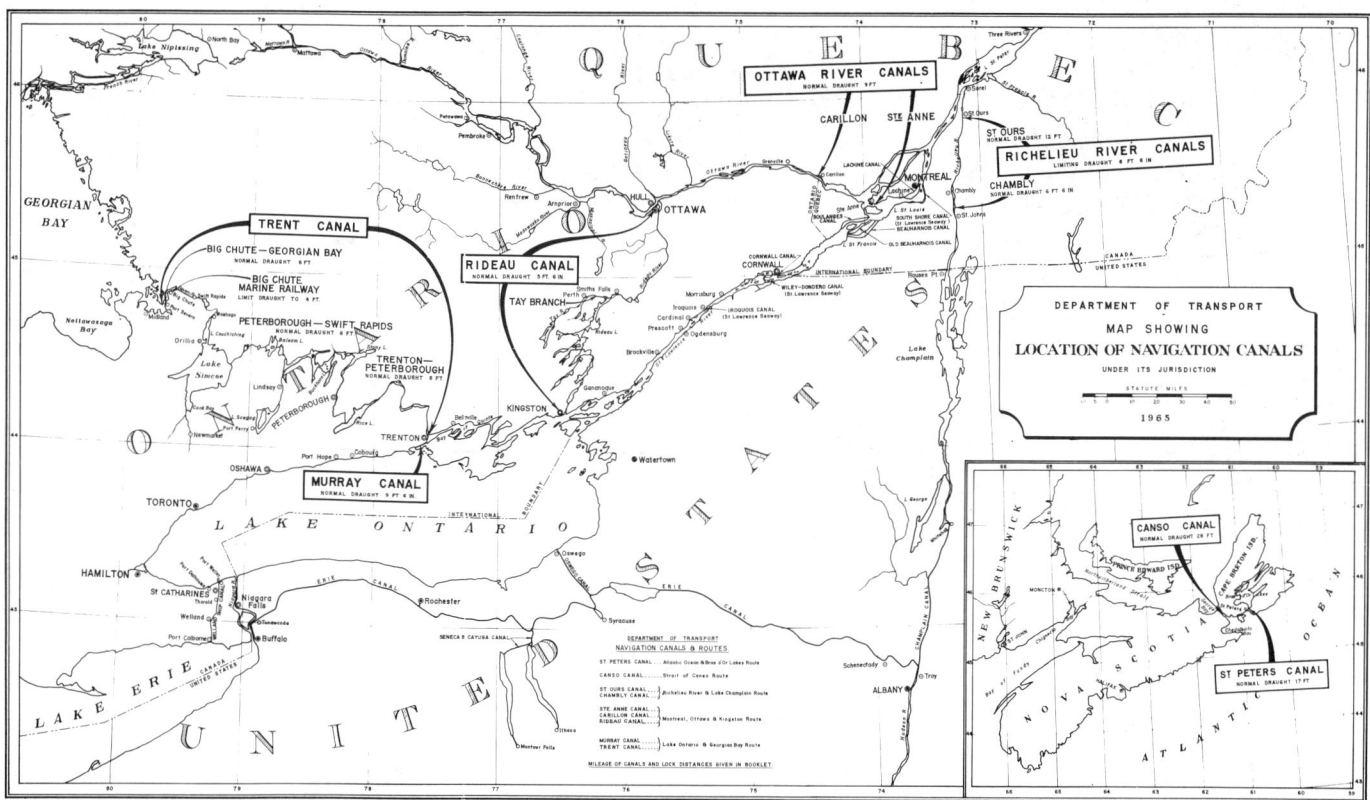

The American Revolutionary War drove thousands of "Tories," loyal to the British Crown, out of northeastern United States and into the Maritime Provinces of New Brunswick and Nova Scotia. These loyal British subjects, along with the French Canadians of Quebec and the British Canadians in the Ontario Province found themselves constantly threatened by the great, growing "Colossus" to the South, a situation underscored by the War of 1812, and subsequent Yankee raids across the St. Lawrence River. Even as late as the American Civil War, when Canada was used as a base of operations for the Confederate spy system against Union troops, anti-American feelings in Canada were apparent.

Canadian citizens, spread thinly along the St. Lawrence River, and the north shores of the Great Lakes, felt a compelling military and economic need to improve their east-west communications. Their only seaports (on the Atlantic) were Quebec and Montreal. The many rapids of the upper St. Lawrence River made water navigation south-west of Montreal impossible — the alternative being a long, overland "Portage" of goods, between Prescott, Ontario and Montreal. At Prescott, water navigation was again possible, to the west, via the Thousand Island section and Lake Ontario, as far as Toronto and Hamilton. Here again the Canadians were forced to "portage" their goods across the Niagara Escarpment to continue water transport along the north shore of Lake Erie, and further west.

The Canadian Government early realized the need for canals to tie their far-flung provinces together. One of their first moves was to build (in 1797) a 300-foot canal at Sault Ste. Marie to bypass the rapids between Lake Huron and Lake Superior. The lock of this canal, one of the first canal locks built in North America, was only 38 feet long, 8 feet 9 inches wide, with a 9-foot lift. This lock was built to fit the long canoes of the early "voyageurs." This lock has today been re-built on the actual spot where it first existed at the Soo. The remaining drop in the river was considered navigable for the boats of the day.

The second canal in Canada came years later, in 1821, when the eight-mile long Lachine Canal was started, to overcome a 47-foot drop between Montreal and

Royal Sappers and Miners, of the 7th and 15th Companies, at work on the Rideau Canal, 1826. From Revue de l'Ingenierie, Mars-Avril 1976. (Courtesy Bob Mayo.)

41

Historic drawing of the flight of locks on the Rideau Canal between Ottawa and the Ottawa River. (Courtesy A. C. Brown and W. D. Naftel.)

exists today, almost exactly as when it was built 150 years ago, and has now become a tourist attraction for boating and canoeing, and hiking and biking along the old Towpath.

Early improvements to navigation of the Ottawa River included a wooden lock (1816) in the western channel at Vaudreuil, and the St. Anne Canal (1840-1843) on the eastern channel, above Montreal. Next came the Carillon Canal (1825-1833) and the Grenville Canal (1825-1829) further upstream. All were less than one-half mile in length and all provided a draft of 6 feet, thus completing the Rideau-Ottawa inland waterway from Lake Ontario to the lower St. Lawrence.

Further west, work had already begun on the greatest Canadian achievement of the Nineteenth Century — the first Welland Canal. The "spark plug" behind this

Lachine, at the mouth of the Ottawa River. The original canal had 7 locks, each 100 feet long and 20 feet wide. It was completed in 1825 and made limited navigation possible between Montreal and Ottawa.

The next step was the building of the Rideau Canal between Kingston and Ottawa in Ontario, the major project of an all-Canadian water route between Lake Ontario and the Atlantic Ocean. Lieutenant Colonel John By of the Royal Engineers was commissioned to build the 122-mile Rideau Canal, a monumental task for the time. Work began in 1826 and was completed in 1832. The original canal included 47 locks (133 feet long by 33 feet wide) and 22 lock stations. There were flights of locks, at a number of points in the route, which wound its leisurely way along the Rideau and Cataraqui Rivers, with a series of dams converting the route into a procession of placid pools. The canal climbs 162 feet from Kingston to the summit level at Upper Rideau Lake and then descends 277 feet, concluding with the magnificent flight of eight locks down to the Ottawa River. Average lift of the locks varies between eight and eleven feet. With a few changes, the Rideau Canal

William Hamilton Merritt, founder of the Welland Canal Company of 1824, and its prime promoter, until completion of the canal in 1829.

enterprise was William Hamilton Merritt, son of a British Loyalist who had left New York State after the Revolution, and settled at the Valley of the Twelve (later known as Saint Catharines) west of Niagara Falls, Ontario. Will Merritt, who became a well-to-do local industrialist, had often thought of the possibility of running boats over the 327-foot high Niagara Escarpment to make a water connection between Lakes Ontario and Erie. This would take the place of the difficult "portage" between the two Lakes, but such an idea seemed like an impossible day-dream. When the British Governor of Ontario placed an embargo on the shipment of Ontario corn into the States, and insisted that it must be shipped

Rideau Canal lock gates are assisted by a hand-operated chain and winch. (Parks Canada photo.)

to England instead, the frustrated farmers and millers of southern Ontario began looking for ways of improving their water transport east and west.

Will Merritt at this time began promoting the idea of an all-Canadian Canal to by-pass Niagara Falls, knowing that it would be difficult and expensive. He made a trip to England to try to raise funds, and enroute succeeded in obtaining American investment money in New York City. The Canadian government also offered financial assistance and finally there seemed to be enough money pledged to get started. The first shovelful of ground was turned November 30, 1824, and after constant pushing and promotion by Will Merritt and his partner, George Keefer, and using up all of Merritt's personal funds, the first Welland Canal began to take shape, with 39 locks, 110' long x 22' wide

An ocean-going vessel passing through one of the St. Catherines Locks on the Welland Canal. (Photo by the Author)

The "Big Chute" carriage dips into the water at the upper level to take on a couple small boats. (Courtesy W. E. Keenan)

with an average lift of 6' to 11'. It was finished in late 1829 and several boats transited the new canal before the winter freeze, even though ice had to be broken away to pass the boats through! The Welland Canal was at last an accomplished fact!

In the meantime, various short canals and improvements had been made to the Saint Lawrence, starting as early as 1779, to by-pass the numerous rapids between Lake Louis and Lake St. Francis. In 1843 work began on the Cornwall Canal to by-pass the Long Sault Rapids and in 1845 the completed Beauharnois Canal provided a route around the Soulanges Rapids above Montreal. When the Williamsburg Canal System was opened between Long Sault and Prescott, the St. Lawrence became navigable its full length for vessels of nine-foot draft, and the need for the overland portage route was eliminated. The original Lachine Canal was also widened and deepened to match the Beauharnois Canal.

As previously reported, the Chambly Canal, started in 1831 and completed in 1943. made a connection from the lower St. Lawrence into the United States via Lake Champlain, and the Champlain Canal to New York City.

The Trent-Severn Waterway is an amazing system of lakes and rivers, running 240 miles across the Province of Ontario between Trenton, on the north shore of Lake Ontario, and Port Severn on Georgian Bay — a part of Lake Huron. It was never a high-priority project for the Canadian Government, and as a result it was under construction (with lengthy delays between active work) over a period of nearly ninety years. It was started in 1833, with the formation of a Commission to make "limited improvements."

A fixed keel boat coming up on the carriage at "Big Chute" (Courtesy W. E. Keenan.)

World's highest hydraulic lift locks on the Trent Canal at Peterborough, Ontario. (Photo by Capt. Tom Hahn.)

It was intended primarily to serve the developing lumber trade in the interior by providing a means of bringing lumber rafts and steam-driven freight boats to market at the ports on Lake Ontario. Its route included dozens of interior lakes, including Lake Simcoe, (the largest) with short channels, locks and dams between lakes.

Between 1833 and 1844, the construction of locks at Glen Ross, Hastings, Whitlaws Rapids, Bobcaygeon and Lindsay was carried out by the Inland Water Commission. However, funds for the project ran out and work was suspended.

The development of new railroads and highways in the area lessened the need for the canal, but recognizing local water traffic needs, the Ontario government some years later authorized locks at Young's Point and Rosedale and a new lock at Lindsay. Between 1883 and 1887 short channels and locks were built by the Federal Government at various points, making navigation possible from Lakefield to Cobonconk and Port Perry. The Peterborough-Lakefield section was begun in 1895 and completed in 1904, with the opening of the celebrated Peterborough double hydraulic lift lock, at 65 feet, the world's highest hydraulic lift lock. There is a similar hydraulic lift lock on the Trent Canal at Kirkfield but with a lift of only 48 feet. An outlet to Lake Ontario came with the completion of the Rice Lake to Bay of Quinte section in 1918.

Work on the Severn Division — from Lake Coughiching to Georgian Bay — was begun in 1914 and completed in 1920. Marine Railways were built at Swift Rapids and Big Chute to carry the boats in special "cradles" between levels at these points. These unusual devices have been improved and up-graded in more recent years. Both "Big Chute" and the Peterborough Lift Locks have become compelling attractions for tourists from all parts of the World.

The first vessel to travel the entire 240-mile waterway was the motor-launch "Irene" which made the trip from Trenton to Port Severn in nine days in July of 1920 — completing a century-old dream!

Of the modern Welland Canals, and the St. Lawrence Seaway, more later.

ILLINOIS AND MICHIGAN CANAL

Perhaps no other canal since the Erie Canal had such a profound effect on the commerce and economics of central United States as the Illinois and Michigan Canal, completed 1836 to 1848.

The land around the southwest curve of Lake Michigan is only eight feet above the level of Lake Michigan, and a few miles further west the drainage area for the Illinois River slopes west and south to join the Mississippi. French explorer Louis Joliet as early as 1673 spoke of the ease with which a water connection could be made between Lake Michigan and the Illinois River, and the desirability of doing so — to link New Orleans with the Great Lakes. The British were never too concerned with the idea in the 1700's, but as the Americans pushed westward into the Indiana and Illinois territories, in the Nineteenth Century, the prospect of such a water connection became the subject of much talk and speculation.

From the time that Illinois officially joined the Union as a State, in 1818, it was assumed that a Canal between Lake Michigan and the Illinois River would be

WILLIAM GOODING
1803 - 1878

William Gooding was born in 1803 in Bristol, New York. He studied engineering on his own. He worked on the Welland Canal in Canada from 1826-1829. He also worked on the Ohio Canal in Scioto until 1832. He surveyed the Erie and Wabash Canal in Indiana from 1833 to 1835. In 1836 he was appointed Chief Engineer for the Illinois and Michigan Canal and held that post until 1848 when the canal was completed. In 1848 he became Secretary to the Canal Trustees. He also served as U.S. Civil Engineer, and Special Commissioner of the Board of Public Works of the City of Chicago. He died in May 1878 and is buried in Lockport.

the prime objective of the new Illinois State Government, and so it was! They immediately petitioned the Federal Government for permission to build the Canal, which was finally granted in the Act of March 30, 1822. At this time no provisions were made for federal financial assistance to the new State on the project. Nevertheless, Illinois formed a Canal Commission the following year which hired Engineers Justus Post and Rene Paul to run preliminary surveys. They came up (in 1824) with five possible routes, and cost estimates ranging from $639,500 to $716,000.

No further action was taken at that time.

In 1827, after repeated appeals from the Governor, the Federal government approved a bill to provide federal lands (300,000 acres) in Illinois which were to be sold to raise funds for the Canal. The Canal Commission was reactivated in 1829 and told to hire an engineer to make more detailed surveys for the Canal route. Engineer James Thompson was retained by the Commission that

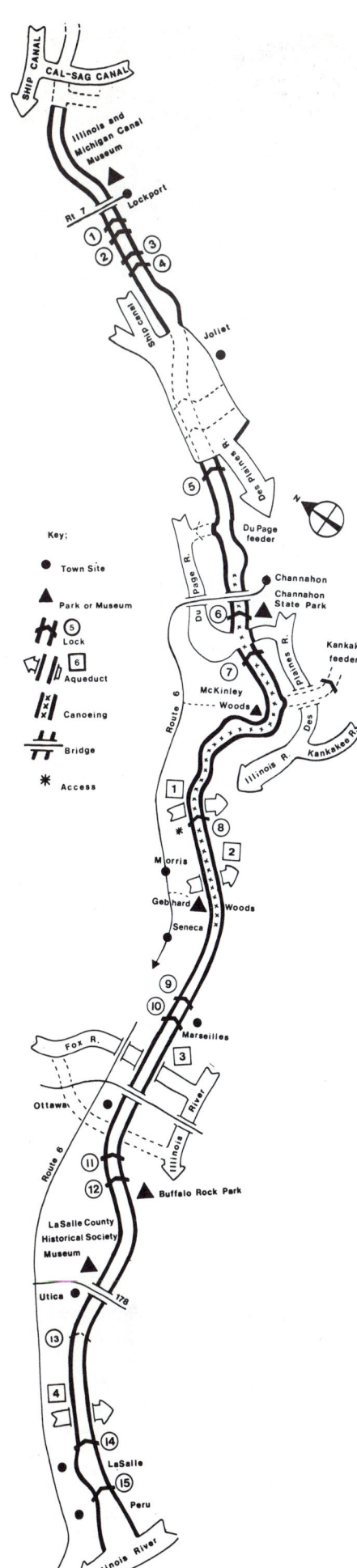

Locks Six and Seven of the original Illinois and Michigan Canal near Channahon, Illinois. (State of Illinois photo.)

same year and made a new survey, with some token help from federal engineers sent in by the War Department. However, little action was taken until James Bucklin in 1830-33 made surveys for both a canal and railroad route.

Bucklin's surveys showed three possibilities: (1) A lake-level canal, cut deeply enough to permit Lake Michigan to feed it, costing $4,107,440.30; (2) a canal with its summit level eight feet above the Lake, costing $1,601,695.83; and (3) a railroad along the same line costing $1,052,488.19. The Commissioners recommended (in 1833) that a railroad was the most logical idea, would be open the year round, would be less expensive to construct, and would be a faster means than the waterway. The governor seconded their report.

However, the supporters of the Canal were still numerous in Illinois and were quite vocal in their repudiation of the Commissioners' recommendation. The election of 1834 became a campaign of "those in favor of the Canal and those against it." The "canalers" won the election "hands down" and Joseph Duncan, a strong believer in the Canal, was elected Governor. His dream was a canal channel deep enough to pass steamboats from river to lake. One of Governor Duncan's strong supporters was a young Springfield attorney named Abraham Lincoln. The legislature in 1835 appointed a new Canal Commission, with power to raise funds and begin work at once.

The sale of lands had not yielded sufficient cash to get things going, so an emmisary was sent East to raise a loan of $500,000 — which was finally obtained, with some difficulty.

The new Canal Commission employed William Gooding, a graduate of the Erie Canal school, as their Chief Engineer, January 9, 1836. Gooding, a man of considerable energy and judgment, remembering the mistake of making the Erie too small, now recommended a canal 60 feet wide at the surface, 36 feet at the bottom and 6 feet deep. On this basis he warned that Bucklin's previous estimate of a lake-level canal was much too low, and re-estimated that plan at $8,654,377. "Canal fever" was in the air, and the Commission recommended that the expensive "lake-level" plan be put into effect at once. Ground was broken, with much ceremony, on July 4, 1836.

From the moment the shovel went into the ground, the size of the town of Chicago began to increase by leaps and bounds! In 1833 Chicago numbered 1200 inhabitants; by 1845 her population had jumped to 12,000; and just before the opening of the canal in 1848, she counted 20,000 inhabitants. By 1854 the figure was 74,500!

For years it had been evident that the Chicago Portage was one of the keys to the continent — the connecting link between the Gulf of Mexico and the Great Lakes. As the news of the Illinois and Michigan Canal construction got around, workmen came from everywhere — from Canada, New England and Ireland to go to work on the I & M Canal. By the end of 1838 there were 2000 men at work.

But all was not sunshine and flowers, during the construction period. In 1837 there was a financial panic which made money scarce and slowed work on the canal to a snail's pace. In 1842 the State

Bank of Illinois failed, and the State itself was close to bankruptcy. Bills of the State bank sold for as little as 38 cents on the dollar in 1842, while Illinois bills sold at auction in Chicago for 18 to 24 cents on the dollar!

Some State legislators were in favor of repudiation of the State's debts, but others argued that the canal must be finished, regardless of the cost, so that it could become a source of revenue which would at least pay the interest on the State's indebtedness. William Gooding was again consulted on the further cost of completion of the original "deep-cut" plan for the canal, which he said would run more than $3,000,000 — a staggering sum for the nearly bankrupt State. However, Gooding pointed out that by reverting to the "shallow-cut" plan, with the summit level twelve feet above Lake Michigan, the cost of completion could be shaved to $1,600,000. It was decided to put the shallow-cut plan into effect; more money was borrowed, and full-scale work resumed in 1845.

As constructed, the I. & M. Canal began at the mouth of the Chicago River, using the river itself as the first "level." It then climbed twelve feet to the Summit Level, which it followed overland to Lockport (the headquarters for the Canal) dropping down to the Des Plaines River Valley through a flight of four locks, and continuing its descent through Joliet, Channahon, Ottawa, via eleven more locks to LaSalle, where it joined the Illinois River. Total length of the canal, from Chicago to LaSalle — 96 miles. Size of the locks — 110 feet long by 18 feet wide, with varying lifts.

The fact that Lake Michigan was not used as a source of water for the entire canal, as originally planned, made it

"City of Pekin" Packet Boat on the Illinois and Michigan Canal, C. 1890 (Courtesy Bruce Anderson)

necessary for Engineer Gooding to provide a number of "feeders." From east to west, these were the Calumet Feeder, the DuPage River Feeder, the Kankakee River Feeder and the Fox River Feeder, providing artificial connections between these rivers and the I. & M. Canal at various levels. Longest of these was the Calumet Feeder, bringing water 17 miles from the Little Calumet River at Blue Island to the I. & M. at Lemont.

With hard times having reduced the cost of labor and supplies — for the first time in history a canal was completed at less than the estimated cost — a further expenditure of only $1,429,606. The entire cost of the canal, not counting interest, had been $6,557,681. Sale of canal lands by 1871 had brought in $5,858,547. so the State of Illinois had almost recovered her original investment at that time.

The Canal was officially opened on April 19, 1848, with a boat starting from each end of the canal, bands playing and champagne corks popping. The celebration in Chicago was particularly hilarious — with good reason. The I. & M. Canal was already making the city prosperous, and would soon make her the "crossroads city" of central USA, with much of American commerce and industry making its headquarters there. A significant event took place a few days after the official opening of the Canal, when the "General Thornton," loaded with sugar from New Orleans, having climbed up the Mississippi and Illinois Rivers, passed through the Canal from LaSalle to Chicago, and on to Buffalo. Nothing like this had ever happened before!

Some of the first materials shipped southward on the Canal were bound for the Mexican War. The Canal Commissioners concentrated on building up Canal traffic to pay off their construction costs, and interest. The old St. Louis overland route eastward, serving Illinois, was replaced by the Chicago water route, giving additional impetus to the growth of Chicago. Previously all traffic had been

Construction (1895) of the Chicago Sanitary and Ship Canal, the training ground for the Panama Canal. (Chicago Historical Society photo.)

handled by wagons pulled by horses and oxen. As the price of transportation rates fell, so did the prices of materials shipped. Lumber dropped from $60 to $30 per thousand board feet; wheat prices dropped from eight to four cents per pound.

$88,000. in tolls were collected from 162 licensed canal boats during the Canal's first year of operation. The trip from Chicago to LaSalle took 20 to 25 hours, averaging four miles per hour over the entire route. Seven thousand boats traveled the Canal in 1862 as traffic continued to increase. The two best years for tolls were 1865 and 1866, during each of which a little over $300,000. was collected. Canal traffic increased over two decades until 1874 when 12,424,705 barrels of wheat and corn were shipped to Chicago. Tonnage reached its peak in 1882 at 1,011,287 tons transiting the Canal. That year only $85,947. in tolls were collected, as rates had been reduced to meet the competition of the railroads.

However, the rapid growth of Chicago as one of the nation's leading grain and meat-packing centers had its problems, the greatest of these being sanitation. For years Chicagoans had been dumping their sewage into the Chicago River, which ran sluggishly into Lake Michigan — the source of their water supply. In 1854, five percent of Chicago's population died of cholera, which brought the City face to face with its water pollution problems.

A set of pumps had been installed where the I. & M. Canal met the Chicago River, to augment the supply of water in the summit level of the canal. It was found that these pumps could be used to pump some of the suspended sewage out of the Chicago River and into the Canal. This led to the idea of using the I. & M. Canal as a means of diverting Chicago's sewerage westward, away from Lake Michigan.

Sanitation problems became so bad in the 1860's that the City of Chicago petitioned the State for permission to revive the old "deep-cut" canal plan of 1836 and use Lake Michigan water to purge the Chicago River of its accumulation of sewage and send it over the divide and down the Canal. The State passed an enabling act in 1865 and work commenced the same year to deepen the summit level of the Canal, with Chicago paying the bill. Construction of the "deep cut" meant the removal of the lock at summit level, south of Romeo. A guard lock was installed at Bridgeport to replace the old lift lock located there. The need for the Bridgeport pumping works was eliminated, as well as the Calumet Feeder, and much later, the DuPage and Kankakee Feeders.

The gigantic forms used in pouring concrete for the 58-foot high lock walls of the Illinois Waterway. This photo was made at Brandon Road Locks, Joliet, Illinois, during construction in 1928. Robert S. Mayo, one of the Engineers for Blaw Knox, the contractor, stands on the concrete at the base of the form. (Courtesy Bob Mayo.)

With this "deep cut" across the divide completed by the City of Chicago in 1871, at a cost of $3,000,000. the upper reaches of the Canal were now deepened to below lake level, permitting Lake Michigan water to flow westward into the Chicago River, sweeping some of the sewage down the canal towards the Illinois River. Because of the City's tremendous losses in the great fire of 1871, the State refunded the money spent by the City in reversing the flow of the Chicago River.

This operation improved Chicago's sanitation picture for the next few years, but a heavy storm in 1885 dumped 5.5 inches of rain into the Chicago area in a 24-hour period, washing much of the accumulated filth of the Chicago River into Lake Michigan, far beyond the city-water intake point. The resulting outbreak of cholera, typhoid and other water-borne diseases killed 12% of Chicago's population.

This disaster hastened action on recommendations by the "Citizen's Association of Chicago" (formed in 1880) who had asked for an entirely new and larger canal, parallel to the I. & M., to carry Chicago's sewage across the divide. In the meantime, the pumping station at Bridgeport had been started up again to help in chasing the polluted water down the Canal. The State finally passed an act in January of 1890 which created the Sanitary District of Chicago, and initiated the construction of the much larger Chicago Sanitary and Ship Canal from Chicago to Joliet, whose prime purpose was to divert Chicago's sewage west and south to the Illinois River. Joliet, already suffering from the unbelievable stench of the sewage arriving there, violently objected to the new canal and other communities as far downstream as St. Louis took up arms against it also. Nevertheless, the Chicago Sanitary and Ship Canal began construction in 1892 and was completed to Joliet in 1901, making connection there with the I. & M. in the "Upper Basin" at Joliet, above Dam Number One. The old I. & M. above Joliet was abandoned as a traffic artery, but continued to handle some local drainage between Lockport and Joliet. Since then the Chicago Sanitary District has developed some of the world's largest inland sewage disposal systems to reduce the pollution of the Chicago and Illinois Rivers.

Many engineers tell us that the Chicago Sanitary and Ship Canal, an extremely wide artificial waterway, became the training ground for the building of the Panama Canal. After World War I, federal funds were applied to the building of the Illinois Waterway, opened in 1933, which supplemented the use of the older canals. Portions of the old I. & M. are now preserved as historic sites.

WISCONSIN'S PORTAGE CANAL

The Steamboat "Swan" is shown on its journey through the Portage Canal in 1896.

(Publisher's Note: We are indebted for the following history to the Portage Canal Society, Portage, Wisconsin, and to Frederica Kleist and Herb O'Hanlon who provided the information for this chapter. We reproduce it here just as originally published in AMERICAN CANALS - Bulletin of the American Canal Society - for February, 1976.)

The name and geographical location of Portage, Wisconsin, have close ties with early American water transportation. Joliet and Marquette, in their famed explorations, were the first to observe that the narrow two-mile strip of land separating the Wisconsin and Fox Rivers, should be connected by a canal. Located on the *Wisconsin River*, Portage, population 7,821, is a short 2½ miles west of the headwaters of the *Fox River*. For almost 100 years waterborne commerce moved up the Fox River from Green Bay and Lake Michigan, crossed the portage and continued down the Wisconsin River to its junction with the Mississippi River.

Here was a logical spot for a canal, and as trade grew, agitation began to cut a channel between the two rivers to open a wider trade route to the west. In 1837 a company was chartered under the name of the *"Portage Canal Company"* to build a canal connecting the *Fox and Wisconsin Rivers*. In 1838, after about $10,000 was spent by the company, the canal was abandoned. Nothing further was done until Congress in 1846, recognizing the value of the route granted the State of Wisconsin alternate sections of land, three miles on each side of the Fox River, to build a canal. A new route was chosen; construction began on 1 June 1849.

The work progressed slowly because of misunderstandings between the contractor and the state. The men working on the canal were not paid for weeks and months and were compelled finally to abandon it in an unfinished condition. A resident of Portage thus describes the work in March 1851: "The banks of the canal are crumbling before the thaw, in many places, and falling into the stream. The planking is in great part afloat . . . It presents a melancholy spectacle of premature decay."

Repairs were subsequently made, the water let in, and on Saturday, 24 May 1851, a boat attempted to pass through the canal. A Portage newspaper describes the scene as follows:

"The beautiful steamer, *John Mitchell*, nearly accomplished the feat of passing through the canal at this place, from the Fox into the Wisconsin River, on Saturday last. She came up as far as Main street. As the John Mitchell came up the canal, the *Enterprise* came up the Wisconsin River to the head of the canal. The blustering rivalry between these inhabitants of different waters (the throat of each giving its best puff and whistle alternately), was quite exhilarating, and called out a large concourse of citizens, to gaze upon the scene presented, and make predictions for the future. After a short time, boats and citizens withdrew, amid strains of music, and the 'noise and confusion' were over."

The water was drawn off, and the work of strengthening the banks and bottom, to prevent the quicksand from pouring in and filling up the bed, was proceeded with; but their efforts were of little avail, if the same local authority quoted can be relied on. On the 31st day of August, 1851, the water was again let in, and the next morning it presented a rather novel appearance, the planking having raised from its fastenings, at the bottom, and floated on either side of the surface, and forming two floating plank-roads. On Sunday night, September 28, 1851, the Wisconsin River, which has been unusually high for some days, broke into the canal, and cut a channel through its southern bank, some fifty yards wide and eight or ten feet deep.

Little was done from this time until 1853. As the Constitution of the State forbade the creation of any public debts, the Board of Public Works was limited in their expenditures to the receipts from the sale of land granted by Congress. At this stage of affairs, another company proposed to take charge of the work, complete the canal and the improvements contemplated on the Fox and Wisconsin Rivers. On the 6th July, 1853 an act was passed by the Legislature of this State, incorporating the *"Fox and Wisconsin River Improvement Company."* The company was instructed to commence the work within 90 days, and to finish the improvement within three years. They failed to comply with the law, and finally the United States took the work off their hands. Then came the Civil War and more delay.

As built by the Government, the canal was commenced in the fall of 1874. The excavation was made by a steam excavator, wheelbarrows and small construction cars. When completed in June 1876, the canal was 2½ miles long, 75' wide and 7' deep, with a timber and pile revetment on each side. Two locks were constructed, one with a lift of 9' at the junction with the Wisconsin River at the western edge of Portage and the second, to the east, near the junction with the Fox River. The latter (Fort Winnebago Lock) had a lift of 6'. The locks were rebuilt by the Government; the former in 1880, and the latter in 1874 and 1875. They were 35' wide and 160' long, between gates. The United States steamer, *Boscobel*, was the first boat to pass through the canal, after its completion.

There was extensive traffic through the canal for 30 years; barges loaded with lead ore from Prairie du Chien, lumber scows from northern Wisconsin and flats of farm produce vied with pleasure craft of all sizes. The canal was closed on 7 July 1951. The locks were removed from the Fox River end but the one on the Wisconsin River end is still intact and can be seen today. It is of modern concrete construction with steel lock gates. Evidently the lock was rebuilt prior to abandonment.

In its passage through the town, the canal is blocked by one small coffer dam and by changes in several street crossings. Bridges have been removed, several lengths of large corregated steel culvert pipes placed in the canal bed parallel to the canal banks, and all filled in to make solid street crossings.

About a mile east of the Wisconsin River the double-track main line of the Milwaukee Road crosses the canal at Portage Junction. An examination of the bridge shows it to be a former vertical-lift span that was made permanent at the close of navigation. Another mile east of the former lift-bridge, and off state highway #33, is a series of old buildings maintained by the State of Wisconsin. A country road from highway #33 to the restored buildings closely follows the waters of Portage Canal. Known as the Indian agent's residence, the buildings, too, are relics of Wisconsin's early heritage.

To the east, are the remains of the downstream lock. Actually only half the lock is still evident. One set of gates with stone lock-walls remains. The gate has been cut down to water level and cut-stone — perhaps from the lock walls — packed into the downstream V of the gate, in effect, forming a small dam. Retracing the canal back to the Wisconsin River shows a waterway still in a remarkable state of good preservation. Canal banks are still held in line by timber cribbing; the water level is intact throughout.

With the exception of New York State, most of these inland waterways in Eastern USA are maintained by the U. S. Army Engineer Corps.

ARMY CORPS OF ENGINEERS

The appointment of Col. Richard Gridley by General George Washington June 16th, 1775 as Chief Engineer of the Continental Army is generally regarded as the act which originated the U.S. Army Corps of Engineers. The initial Corps assembled by Gridley consisted of a few frontier surveyors and a handful of trained Frenchmen. It was dissolved in 1783. In 1794 President Washington appointed a Frenchman, Etienne Rochefontaine, to command a new Corps, established by Congress and called the "Corps of Artillerists and Engineers." A year later, the training of young officers and men in the art of military engineering began at West Point, New York.

The modern Corps of Engineers was created March 16, 1802 when President Thomas Jefferson was authorized by Congress to establish a corps of five officers and ten cadets, who "shall be stationed at West Point, in the State of New York, and shall constitute a Military Academy; and at all times to do duty in such places and on such services as the President of the United States shall direct." The Military Academy, influenced by French instructors and textbooks, was the first school of engineering in the United States.

Since the early days of steamboat navigation in the USA, various individual entrepreneurs, such as Robert Fulton and his partners, had obtained exclusive navigation rights to various rivers from the States through which they ran. It was not until 1824, during the test case of "Gibbons versus Ogden," that the national legislature was given full authority over "navigation within the limits of every state of the Union." Even though there were private companies operating navigational enterprises in various rivers well into the nineteenth and twentieth centuries, the Corps of Engineers, acting for the federal government, was empowered to acquire such facilities and improve them, if such action seemed to be in the best public interest.

In 1824 Congress authorized the President to make "surveys, plans and estimates for the routes of such roads and canals as he may deem of national importance, in a commercial or military point of view, or necessary to the transportation of the public mail." and the employment of "two or more skilled civil engineers, and such officers of the Corps of Engineers as he may think proper."

Another authorization in 1824 provided $75,000 for improvement of navigation of the Ohio and Mississippi Rivers and the employment of "any of the engineers in the public service which the President may deem proper." Such authorizations, repeated by almost every Congress since 1824, have become known as the Rivers and Harbors Acts. Through the years the responsibility of the Corps of Engineers had been extended to flood control and multi-purpose control of the river basins of the Nation for optimum use and conservation of all water facilities.

In addition to its complement of Army Engineer Officers, the Corps currently employs some 40,000 civilians and operates a fleet of 3000 vessels such as dredges, barges, pontoons and other floating equipment in connection with their river and harbor maintenance and operational activities.

ST. MARYS FALLS CANAL

The sixty-mile long St. Marys River is the only connection between Lake Superior and Lake Huron. There is a 23-foot differential in elevation between Lakes (Superior the higher) most of which is concentrated in a half-mile section of rapids on St. Marys River at Sault Ste. Marie, Michigan, or Sault Ste. Marie, Ontario — depending upon whether you approach the River from the American or Canadian side. Americans refer to the area simply as "The Soo."

As previously noted, in our chapter on the Canadian Canals, the first attempt at an artificial waterway in Canada was a short canal and lock — one of the oldest in North America, built on the Canadian side of the Soo in 1797. Prior to this all furs and goods passing between Lakes Superior and Huron had to be "portaged" by land around the white water at Sault Ste. Marie. During the War of 1812, the Canadian lock was destroyed by the Americans, so traders again had to portage their goods around the falls, for the next forty years. The Americans gained control of the south side of the River in 1820 with the "Treaty of the Sault," signed with the area Indians.

In 1853, after long prodding by Senators Lewis Cass and Alpheus Felch of Michigan, the U. S. Congress granted the State of Michigan the right to construct a canal at the Soo, and gave them 750,000 acres of public lands which could be sold to finance the project.

An agreement between the State of Michigan and the Fairbanks Scale Company, which had extensive mining interests in the Upper Michigan Peninsula, was signed April 5, 1853 for construction of the canal and locks. Charles T. Harvey, Fairbanks General Agent, was put in charge of the project. The agreement specified that the canal must be operational in two years or Fairbanks would receive nothing for its efforts.

Ground was broken June 4, 1853. 400 men were on the first payroll, but the labor force soon swelled to 1600. Under summer sun and biting winter cold, Harvey drove his men to meet the completion deadline. Fairbanks Company agents met ships at eastern seaports to recruit immigrant laborers to replace the vacancies in Harvey's labor force created by men quitting and by cholera.

The job was done on time, and turned over to the State, with two 350-foot long, 70-foot wide locks in tandem, each with a lift of approximately ten feet. This combination was known as "State Lock." On June 18, 1855, the Governor of Michigan and other dignitaries assembled at the Sault to watch the steamer "Illinois" pass through the canal and its locks between Lakes Huron and Superior. Heavy ship traffic began almost immediately.

By the 1870's it was obvious that additional locking capacity was needed,

Colonel Orlando M. Poe was a Topographical Engineer before organizing the Ohio Volunteers and serving in the Civil War under General William T. Sherman. In 1883 he was made Superintending Engineer for the improvement of rivers and harbors in the Lake Superior and Huron area. (U.S. Corps of Engineers photo)

beyond the financial limits of the State of Michigan. At this point the Federal Government took over the operation, which was placed in the hands of the U. S. Corps of Engineers. The Corps immediately began construction of a second, single-stage lock, named the "Weitzel Lock" in honor of the District Engineer in charge. This lock (opened in 1881) was 515-feet long by 80-feet wide, and was located parallel to, and to the south of, State Lock. Operation of all locks was made toll-free.

In the meantime, Canada got into the act. In 1887 the Canadian Government let contracts for a 3-1/2 mile long canal on their side of the Sault Rapids, with a single lock 900 feet long, 60 feet wide with a depth of 20 feet, 3 inches over the sills. This system was opened for navigation in 1895 to handle traffic from the steadily growing mineral and timber industries along the north shore of Lake Superior. The Canadians were astounded at the tremendous volume of traffic which immediately began flowing through their canal.

Both Canadians and Americans discovered that they could harness the Sault Falls to generate power, and did so as soon as hydro-electric power generating equipment had been perfected. Thus the entire flow of the St. Marys River was eventually controlled, either for power, or navigation.

The American Locks at the Soo. Note power house at the upper left. (Courtesy Michigan Travel Bureau.)

Even with the opening of the Canadian canal, traffic continued to out-grow the capacity of locks at the Soo. Ships passing through the locks also grew in size. In 1896 the original "State Lock" was replaced by the Poe Lock, named in honor of Colonel Orlando M. Poe, a Civil War veteran, who was made Superintending Engineer of the Soo Locks in 1883. This lock measured 800 feet long by 100 feet wide. The next major change took place when two new locks, the Davis (1914) and the Sabin (1919) were opened side by side, bringing the total number of American Locks at the Soo to four. The Davis and Sabin Locks, oldest locks still operating, are identical, measuring 1350 feet long, 80 feet wide and 23.1 feet deep.

More recently, the Weitzel Lock has been replaced by the MacArthur Lock (1943) measuring 800 feet long, by 80 feet wide by 31 feet deep; and the old Poe Lock has been replaced (1968) by the New Poe Lock, a gigantic 1200 feet in length by 110 feet wide by 32 feet deep.

As an extension of the St. Lawrence Seaway, the Soo Locks today have become the busiest waterway system in the entire world! Ships of all nations pass through the Soo, carrying wheat, iron ore and other commodities east from the Mesabi Iron Range and the great central wheat fields of Canada and the United States. In 1971 the Soo reported 91.5 million tons passing through the American Locks and 1.5 million tons through the Canadian Canal.

A large freighter emerges from Poe Lock at the "Soo".

THE PANAMA CANAL

The opening of the Panama Canal in 1914 was the fruition of a dream of the Spanish Conquistadors exactly four hundred years earlier, and the completion of the most significant engineering, commercial and military achievement in the History of Mankind.

Since Vasco Nunes de Balboa first set eyes on the Pacific Ocean, after crossing the Panamanian Isthmus in 1513, men had dreamed of a water connection here between the Atlantic and Pacific which would immeasurably shorten the long journey around Cape Horn on the tip of South America. First the Spanish used an isthmus crossing to consolidate their conquests in Central and South America, searching without success for a river which might make a connection between the seas. In 1630 a band of English Puritans gained a brief toe-hold on islands off the coast of Panama, only to be driven into the sea by the Spanish eleven years later. Sir Henry Morgan and his Buccaneers soon afterward attacked Panama, made friends with the Indians and investigated a possible route across the Isthmus via the San Juan River to Lake Nicaragua. Other English expeditions over the next century or more attempted unsuccessfully to colonize Nicaragua as the best overland route to the Pacific.

Spanish control and influence in the Americas gradually came to an end in the early 1800's, and the Dutch, Germans and British sent exploratory parties into the interior of Central America with the idea of building a canal between the two oceans. A German geologist, Alexander von Humboldt, after a visit to the area, wrote a treatise entitled "Political Essay on New Spain" which discussed the many international advantages of a canal across Central America and traced nine possible routes, including Tehuantepec, Nicaragua, Panama and Darien.

The great German writer, Johan Wolfgang von Goeth (author of "Faust") read Humboldt's Essay about 1825 and made these comments: "If a project of this kind succeeds in enabling ships of all sizes and lading to go through a canal from the Gulf of Mexico to the Pacific Ocean, incalculable benefits will accrue to the whole of civilized and uncivilized humanity.

"It would surprise me, however, if the United States allowed such an opportunity to slip from their hands. We may expect this youthful power, which already possesses a tendency to move westward, to occupy and settle the vast areas of land beyond the Rocky Mountains in thirty or forty years. Further, we may expect that along the whole length of the Pacific Coast where nature has already provided them with the largest and safest harbors, there will soon arise commercial cities of the utmost importance, with trade flowing out from the United States to China and the East Indies. And if this happens, then it becomes desirable and almost necessary that merchantmen and warships should have rapid passage between the east and west coasts of North America — much easier than the wearisome, disagreeable and expensive journey around the Cape.

"I repeat then, it is absolutely indispensible for the United States to effect a passage from the Gulf of Mexico to the Pacific Ocean, and I am certain they will do it."

However, the United States appeared uninterested in such a venture until the War with Mexico gave them California in 1848, with the almost immediate discovery of gold in that new territory. This led to a rush of fortune-seekers from the east coast to the west coast, and most travelers sought a sea route rather than the long and arduous route across the western plains. Ship service was established to Panama from both east and west coasts, where only fifty miles of land travel over the Isthmus was necessary; but difficult.

The need for a better means of transport across the Panamanian Isthmus became immediately apparent. A railroad appeared the best short-term solution and American interests in New York State sponsored the formation of the Panama Railroad Company in 1849. This railroad was built across the swamps and bogs in the tropical wilderness of Panama with Chinese as well as black and Irish labor, many of whom sickened and died on the job as a result of yellow fever and typhoid. The Americans were not immune to the tremendous hardships of the route or to the local diseases and many of them perished before the job was finished and the first locomotive chugged its way over the forty-five mile route between the seas in 1855. The railway had cost $8,000,000 and 835 lives.

As a result of a dispute over the use of Nicaragua as a canal site, the United States and Great Britain in 1850 entered into an agreement for the avowed purpose of hastening the construction of a ship canal across the Isthmus, known as the Clayton-Bulwer Treaty. This treaty indicated that they would give "their support and encouragement" to all efforts in that direction, and would extend their joint protection over any canal or railway that might be constructed. They agreed that neither of them would exercise exclusive control over an Isthmian Canal, but that they would mutually guarantee its neu-

Theodore Roosevelt, one of the most vigorous Presidents to ever occupy the White House, championed the Panama Canal. He overcame tremendous international and political problems to make it a reality. (Photo, Compton's Pictured Encyclopedia, 1954.)

trality and security and would invite other nations to co-operate in protecting it.

Essentially this agreement blocked the United States from taking the initiative in the building of the canal, for the next fifty years, but opened the way for a third power — France — to do so.

FRENCH CANAL

With the completion of the Suez Canal (1859-1869) Ferdinand de Lesseps, of France, the architect of this sea-level canal, became a sort of international hero, and the world "expert" on canal building. When, in 1878, French diplomats obtained a charter from the Columbian government to build a Panama Canal, the name "de Lesseps" was on everyone's lips to lead the enterprise. And so he did. At age seventy-five — and with unfailing confidence, de Lesseps announced that there would be no great problem building a sea-level canal across Panama, just as he had done across Egypt. Engineers, sent later to examine the route of the canal, recommended that a lock-canal might be better, but de Lesseps brushed all these reports aside and insisted that a sea-level canal was the only way of doing the job. In 1879, La Compagnie Universelle du Canal Interoceanique de Panama was formed, more generally known in America as the "Panama Canal Company." Company stock was sold primarily in France.

United States, ham-strung by its treaty with Great Britain, and anxious to see a canal built, cooperated by offering to sell the French the Panama Railroad as an important instrument in building the canal — at a figure of $14,000,000. De Lesseps, perhaps thinking he could get a better price, declined the offer, but the following year, after the stock of the railroad had climbed from $200 to $250 per share, changed his mind. By this time however, the American interests had decided to retain 25% of the stock to keep the charter alive, selling the Panama Canal Company the other 75%.

The first work force arrived in Panama in January of 1881, but due to political manipulations and contractor problems (which characterized the entire operation) actual construction did not begin until 1882, and on an increasing scale in 1883, 1884 and 1885. De Lesseps appears to have spent much of his time the first several years (when he was presumably supervising construction) in triumphal meetings with local officials in Panama and the United States, and in public relations campaigns to try to sell his company stock abroad.

As with the Panama Railroad, the French Company had not anticipated the tremendous difficulty of attempting to maintain a vast labor force in the jungles and swamps of Panama without extraordinary attention to sanitation and health problems. Nearly half their laborers developed yellow fever, malaria, or other tropical diseases. Many of them died. They were further plagued by an insurrection in March of 1885, during which the city of Colon was burned to the ground and the city of Panama threatened. Order was restored only on the intervention of President Cleveland of the United States, who sent three war ships and five hundred marines, thus saving the city of Panama from destruction, and further damage to canal property — not to mention serious interruption to the construction work.

Early in 1885 it became apparent that the Panama Canal Company was in financial difficulty. The cost of the work had grown considerably beyond the estimates, and new investment money was not available. De Lesseps asked the French government for permission to issue lottery bonds for a loan of $210,000,000 and then left on a triumphal campaign through Panama to renew his image as the world's foremost canal-builder. French wine and a series of dinners for the local dignitaries kept them convinced that all was well. De Lesseps' dramatic performances apparently convinced the populace at home that the project was still progressing and they poured additional funds into stock purchases. However, general mis-management, unbelievable extravagances in the purchase of local facilities — not to mention political manipulations in Paris, were all at work to bring the Panama Canal Company down to ignominious failure in several more years. Notwithstanding all this, a great deal of meaningful work was actually completed by the courageous French engineers and their work forces, under almost impossible working conditions. By 1888, the Company had spent $262,684,000, had excavated 70,565,793 cubic yards of earth, had lost some 22,000 workers' and principals' lives in the process, had run out of funds and had gone bankrupt. It had become an international disaster which shook the entire financial world and was a matter of intense embarrassment to the French government for years.

A new canal company was formed by the receivers in 1889, who were finally obliged to suspend work on the project in May of that year, and whose later efforts were devoted to trying to sell what was left of their assets to the United States.

THEODORE ROOSEVELT 1858 - 1919

Like De Witt Clinton with the Erie Canal, Theodore Roosevelt was the man most responsible for pushing the Panama Canal through to completion. In his position as President of the United States, he had the necessary power to do so, and he did not hesitate to exercise it, in the right places and at the right time.

As the Twenty-Sixth President of the United States, who moved into the presidency September 14th, 1901 as a result of the assassination of President William McKinley, "Teddy" Roosevelt was one of the youngest presidents (43) and certainly the most colorful personality

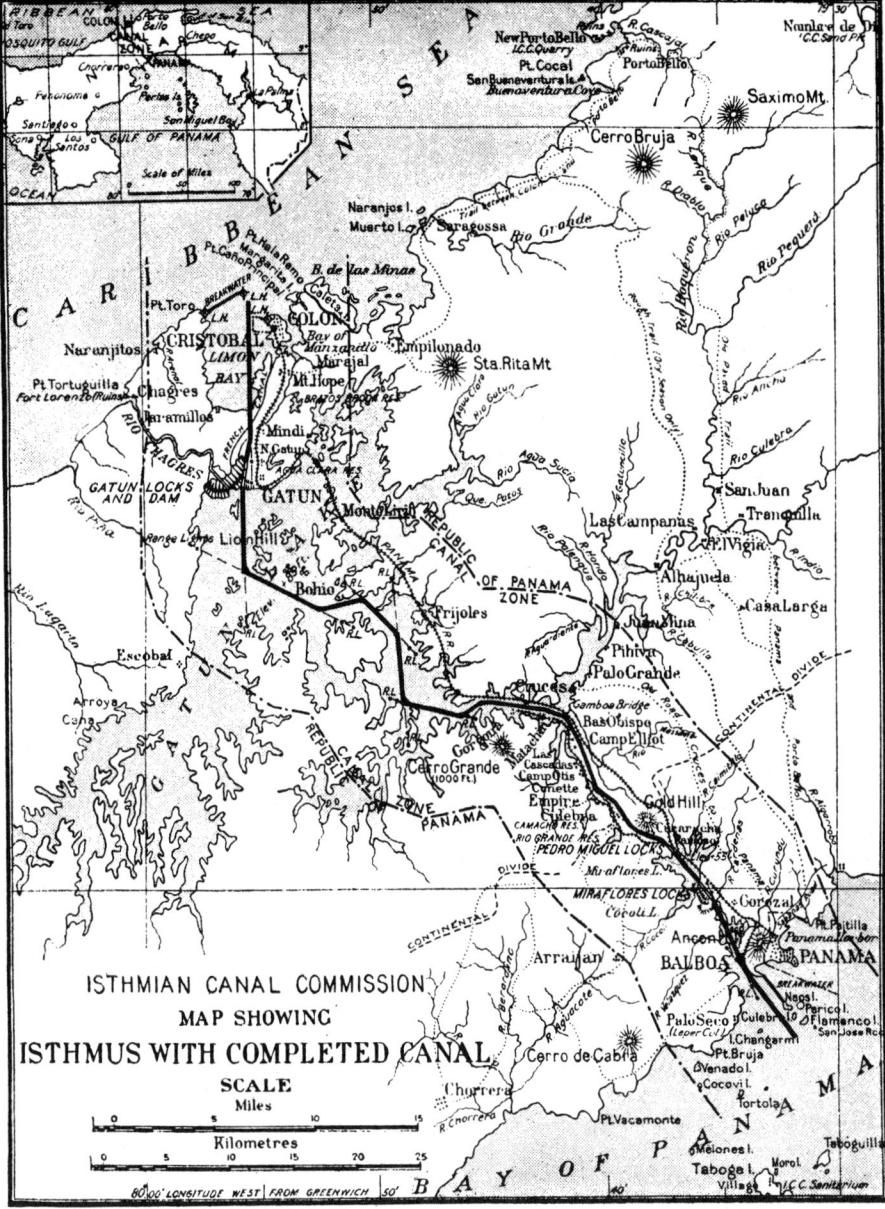

to occupy the "White House" (his own name for it) in the previous history of the United States.

Rancher, hunter, explorer, author, former Mayor of New York, former Governor of New York State, former under-secretary of the U. S. Navy, former Vice President and hero of the Spanish-American War (he led the famous charge up Kettle Hill in Cuba in 1898), the new President brought with him a life-style and a reputation for "getting things done" which had not been seen in Washington since the days of Andrew Jackson. He was the ideal man to cut national and international political "red tape" and get the long-debated Canal underway.

Ever since the failure of the French canal it was becoming increasingly obvious that if an Isthmian Canal was ever to be built, it would have to be done by the United States.

Culebra cut on January 4, 1913, just after a land slide. Note huge track-mounted steam shovels at the right. ("The Panama Gateway," Joseph Bucklin Bishop, 1913)

Construction of the Pedro Miguel Locks. Tremendous concrete "pourings" were required. ("The Story of the Panama Canal," Logan Marshall, 1913)

A private firm, the Maritime Canal Company, had begun work on a Nicaraguan Canal in 1890, and had spent $6,000,000 on actual construction work, which in 1892 exhausted their funds. There were many in Congress who felt that the U. S. Government should have supported and encouraged this enterprise, but the Clayton-Bulwer Treaty still prevented official recognition of the operation.

The Spanish-American War of 1898 made it crystal-clear that a canal across Central America was a war-time necessity. The Warship "Oregon," caught in the Pacific during the blockade of Cuba, made a full-steam run around from San Francisco through the Straights of Magellan to Key West, which took from March 19th to May 26th. The sinking of the Spanish fleet had to be postponed until the "Oregon" arrived!

Shortly after Roosevelt took office, and after considerable wrangling between the British and Americans, a treaty was finally negotiated between Secretary of State John Hay and the British Ambassador to the United States, Lord Pauncefote, and ratified in December of 1901, which specifically abrogated the Clayton-Bulwer Treaty. The Hay-Pauncefote Treaty gave the United States the right to exclusive ownership of an Isthmian Canal, permitting fortification of the Canal and its approaches, and omitting the former requirement that the canal should be kept open to all nations in both time of war and peace.

In anticipation of the signing of the new treaty, Congress had been arguing for years, not about the building of the canal, but about the route to be followed. There was extremely strong support for a Nicaraguan Canal, using some of the work already done by the Maritime Canal Company. Even though the Panama route had been discredited by the failure of the French, there were still those in Congress who argued strongly that the holdings of the old French company, now known as the "New Panama Canal Company" should be acquired by the United States, to salvage what work had already been done there. Debate continued, often with great animosity, between the two factions, until economic considerations and the action of President Roosevelt, resolved the issue in favor of the Panamanian route.

Details of the final decision were as follows: On March 3, 1899, President McKinley had appointed a Commission of nine members, headed by Rear-Admiral J. G. Walker, to investigate all possible canal routes across the Isthmus — particularly those at Nicaragua and Panama — to be "under the control, management and ownership of the United States." The commission estimated that a Canal at Nicaragua would cost $189,864,062 and a Canal at Panama, $144,233,000. The Commission also negotiated with the New Panama Canal Company in France for acquisition of all rights, franchises and property of their partially-completed route in Panama, for a consideration of $109,141,500 — bringing the cost of a Panama Canal route to a total of $253,374,500.

While the Commission felt that the real worth of the French holdings was only $40,000,000, the price that the French were asking ruled out the Panama route, and they therefore recommended, in their report of November 16, 1901, that the "most practical and feasible route" was by way of Nicaragua. This report was accepted by President Roosevelt and transmitted to Congress on December 4, 1901. On January 8, 1902, the House of Representatives voted 225 to 25 to authorize the President to proceed with the Nicaraguan Canal at a cost of $180,000,000, appropriating $10,000,000, for immediate use.

In the meantime, the French company, seeing its last chance slipping away, cabled Rear-Admiral Walker that they would sell at his estimated valuation figure of $40,000,000. On January 18, 1902 the Walker Commission gave President Roosevelt a supplementary report, and the recommendation (in view of the changed conditions) that Panama was now the "most practical and feasible route." This new report threw Congress into complete confusion.

By this time the Nicaraguan Bill had gone to the Senate, and after long and heated debate, and an amendment by Senator Spooner changing the route to Panama, passed the Senate by a vote of 67 to 6 on June 19. It then went back to the House where it passed on June 26th by a vote of 259 to 8. The Spooner Bill was signed into law by President Roosevelt on June 28, 1902, committing the United States to build a canal through Panama, after first acquiring the holdings of the new (French) Panama Canal Company.

Next came negotiations with the Columbian government, conducted by Secretary of State John Hay, offering that government $10,000,000 in gold initially and $250,000 in gold annual rental after nine years (the estimated time for construction of the canal). The Columbian government, holding out for more money, refused the proposal in August of 1903. The State of Panama, urgently desiring the Canal, had threatened to revolt from the Columbian government if they did not accept the proposal of the United States. Understandably, President Roosevelt did nothing to prevent such a development — as a matter of fact sent warships to Panama to protect the American consulate there during the impending revolution.

Under the watchful eye of the American naval forces, the Panamanian Revolution took place as a sort of political formality on November 4, 1903, with only one casualty, when a Columbian ship fired on Colon, killing a Chinaman. The Republic of Panama was officially recognized by Secretary John Hay on November 6, 1903. A subsequent treaty with Panama was approved by President Roosevelt on February 26, 1904, guaranteeing the independence of the Republic of Panama and paying them $10,000,000 initially and $250,000 annually beginning nine years later. In return, Panama granted the United States, in perpetuity, a strip of land ten miles wide, and extending three marine miles into the oceans at either terminal, for the purpose of building and maintaining the new canal.

WILLIAM CRAWFORD GORGAS 1854 - 1920

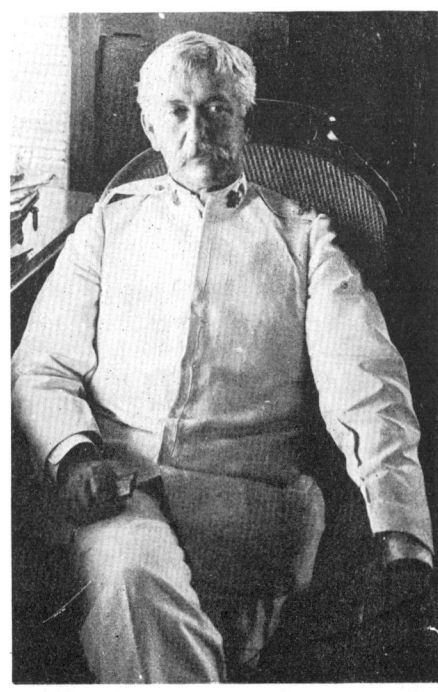

Colonel William C. Gorgas, the man who made the Canal Zone safe for human habitation. ("Panama and the Canal," Willis J. Abbot, 1913.)

The first consideration in the building of the Panama Canal was the health of the vast army of workers which would be shortly laboring in the steaming jungle of inland Panama, with all its tropical diseases — yellow fever and malaria having proven the worst enemies of workers there in the past, on both the railroad and the French canal. The man selected to lead the fight against the jungle diseases was Dr. William Crawford Gorgas, as Chief Sanitary Officer.

Gorgas' father, Josiah Gorgas, had been a General in the Confederate Army and his son had vivid memories of his mother and family leaving their burned-out home in Richmond near the close of the War and moving to Baltimore, with nothing but the clothes on their backs, while his father went south with Lee's army. Young Gorgas received his education at the University of the South in Sewanee, Tennessee, and Bellevue Medical College in New York, entering the army as a surgeon. In his first assignment, at Fort Brown, Texas he became interested in yellow fever, which he was later to combat and conquer.

During the Spanish-American War, Dr. Gorgas served as Chief Sanitary Officer in Havana, Cuba — for years a notorious center for yellow fever. Here he earned world fame by virtually ridding the city of this disease. He was the first to apply the discoveries of the English Army Surgeon, Dr. Ronald Rose, that malaria is conveyed by the bite of the Anapheles mosquito, and those of Dr. Walter Reed, a surgeon in the U. S. Army, that yellow fever is passed from man to man by the Aedes mosquito. Gorgas' technique was the elimination of the breeding grounds for both types of mosquito, no matter how costly such an operation might be.

In Panama, Dr. Gorgas drained every lake, swamp, pond and ditch that could be drained. Over those that could not be drained, he spread a film of crude petroleum to destroy mosquito eggs and larvae. Within a radius of a hundred yards of all human dwellings, he kept the jungle grass cut to ground level, destroyed all rubbish, rats and vermin. Gorgas raised all buildings on stilts above ground level, screened-in porches, windows and doors and ordered all the inhabitants, to keep vessels of water covered when not in use. He built hospitals for isolation and treatment of any victims of the dread jungle diseases. On each train crossing Panama a medical car was included. City water supplies were cleaned up and sewers were dug. Never had there been such a thorough "purge" of all possible conditions which might lead to disease or death for any of the Americans and other workers on the Canal. By the time the Canal was opened in 1914, Gorgas had reduced the death rate of the 39,000 employees in the canal zone to a mere 17 per thousand, from all causes — lower than the normal death rate in most American cities of the period.

Without the work of Dr. Gorgas, the building of the Panama Canal would not have been possible. Already a Colonel by act of Congress, for his work in Havana, Dr. Gorgas was promoted to Surgeon General of the U. S. Army in 1914, and the following year was made Major General.

THE FIRST CHIEF ENGINEERS

John Findley Wallace was selected as the first Chief Engineer of the Panama Canal project, under the watchful eye of a Canal Commission of seven in Washington, who carefully controlled delivery and costs on all materials bound for Panama. Wallace complained later of endless "red tape" and "lack of a free hand" as he had been promised, but actually he was not the man for such a colossal job. To begin with, he hated life in Panama, was deathly afraid of contracting one of the tropical diseases, and never really was able to formulate an effective plan of attack for building the canal. He dabbled with repairs to the machinery left behind by the French, supervised some of the excavation work

Spillway of Gatun Dam, Regulating weirs had just been positioned between the piers when this photo was made, June 1, 1913. ("The Panama Gateway," Joseph Bucklin Bishop, 1913)

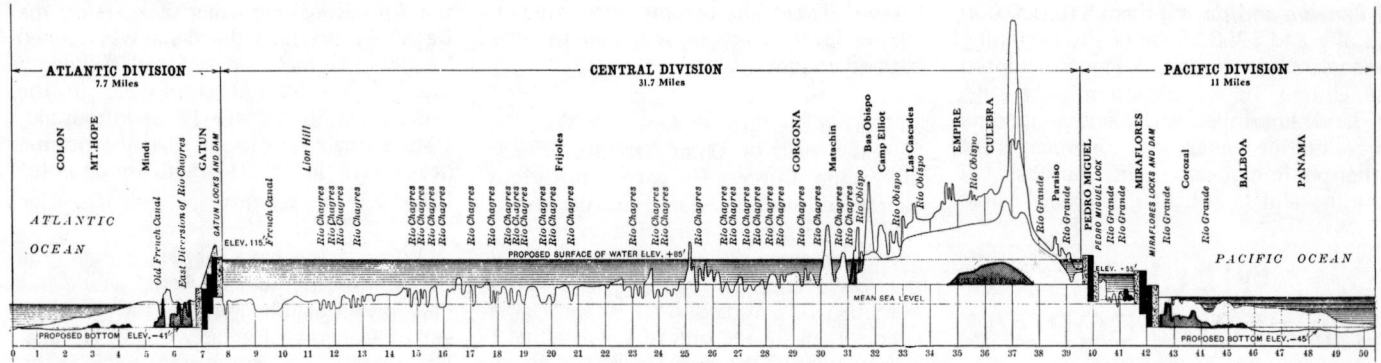

A profile of the Panama Canal, showing the location of the six locking sequences, the Gatun Lake level, the excavation necessary through the Culebra section, the small lake section at Miraflores, and the additional excavation necessary in the Pacific channel to provide an average channel depth of 45-feet. (From "Panama and the Canal" by Willis J. Abbot, 1913)

at Culebra Cut and complained constantly to "Boss" William Howard Taft, Secretary of War under Roosevelt, that he was being ham-strung by the Commission and the endless paper work of obtaining necessary supplies for the job site. Taft did what he could to simplify the procurement procedures and finally even cut the size of the Commission to three people. However, when Wallace came to Washington to say that he could find a better job elsewhere, Taft promptly fired him.

Taft, whom Roosevelt had now placed in charge of the Panama Project, then hired another civilian engineer, John Stevens, in July of 1905. Stevens, primarily a railroad engineer, proved to be a brilliant addition to the project and got things moving almost at once. He brought in some of the most modern earth-moving machinery in the form of giant steam shovels, recently developed in the USA. He rebuilt whole sections of the Panama Railroad, and imported larger rail cars to carry dirt and rock away from the digging sites as rapidly as the steam shovels could load them.

He took full advantage of the partially completed work of the French. Stevens recognized at once the importance of the work that Dr. Gorgas was doing, and gave him every possible assistance in the development of new sanitation facilities and insect-free housing for the inhabitants of the Canal Zone.

To this point no decision had been made as to whether the Canal was to be a sea-level cut, or a lock-canal. A group of engineers, some of whom had worked on the Soo Locks, were sent down to Panama, and they convinced Stevens that a lock-canal was necessary. Stevens became the most ardent proponent of a lock-canal at later Congressional hearings on the matter. Stevens was a bundle of energy and seemed to be everywhere, directing the work in a competent and efficient manner. Everything seemed to be going well.

Then suddenly, on January 30, 1907, Stevens (for reasons still not clear) wrote a letter directly to President Roosevelt saying, in effect, that he was not really interested in the entire project, did not really like the work, and wished that he had taken a position back in the States instead. While the letter was not actually a resignation, Roosevelt took it as such, and wired Stevens that his resignation was accepted.

GEORGE WASHINGTON GOETHALS 1858 - 1928

On the recommendation of William Howard Taft, Army Major George Goethals was summoned to the White House and told that he was to go to Panama to replace John Stevens as Chief Engineer and was also to be made Chairman of the Canal Commission, with full authority over all affairs in the Canal Zone.

Born in Brooklyn, N.Y. in 1858, Goethals worked his way through three years at City College of New York and entered West Point, where he was graduated in 1880 and chose to serve in the Army Corps of Engineers. His career had included work on improvements to navigation on the Ohio, Cumberland and Tennessee Rivers. On the Muscle Shoals Canal (1889-1894) he designed and built a lock with a record lift of 26 feet. When he came to Taft's attention he was a member of the Army Chief of Staff Corps, specializing in coastal defenses. He had already earned a reputation as an expert engineer and an inspiring leader of men.

Colonel (later General) Goethals went to Panama and took over Stevens' duties, effective March 31, 1907. It was now apparent that the Army had taken over, and a number of civilian staff members left, shortly after Stevens' departure. Goethals made a number of changes to his general staff, insisting on complete loyalty to the project as the basic requirement.

A typical lock-gate operating mechanism. The large "Bull Wheel" gate at the left is electrically driven to actuate the arms which open and close the gates, with extreme precision. ("The Story of the Panama Canal," Logan Marshall, 1913)

Colonel George W. Goethals, the undisputed "boss" of all operations in the Canal Zone, after 1907. ("The Story of the Panama Canal," Logan Marshall, 1913)

In his capacity as both Chief Engineer and General Superintendent of all Canal Zone affairs, Goethals was first regarded as a "cold fish," but he soon gained the respect of his subordinates and the populace, as a man who was willing to hear all sides of a question, and who was extremely fair and impartial in his judgments. He divided his time between the office and the actual construction sites, touring the Zone on a special motor-driven car that ran on railroad tracks — which his men nicknamed "the Brain Wagon." By driving himself and his men he completed the canal one year ahead of schedule.

And so, the greatest construction project of all time was carried through to completion by the Army of the United States, under the guidance of the man who became known as the "Benevolent Despot" of the Panama Canal — the undisputed "boss" of the entire operation, in all its ramifications, responsible only to the President of the United States.

CONSTRUCTION

After it had been decided that the canal was to be a lock-canal rather than a sea-level canal, an artificial lake was planned as part of the summit level — Gatun Lake. To create this lake, the world's largest dam was built across the Chagres River at the Carribbean end. The Lake was then extended across the Continental Divide at Culebra with a "cut" at a much higher level than would have been necessary with the sea-level plan. This meant much shallower excavation. In view of all the difficulties with land slides which later developed at Culebra this turned out to be a wise decision.

Ships were to be raised to the Lake Gatun level, 85 feet above sea level, by three lock-sequences at both the Atlantic and Pacific end of the summit. The first locks were a flight of three passing the Gatun-Lake Dam; the next was single — at Pedro Miguel; and the last, a two-flight lock sequence at Miraflores — to drop the ships to the Pacific Ocean level. Because of the heavy traffic which was anticipated, it was decided early in the planning, to make all locks double and all of them identical in size — 1000 feet long by 110 feet wide. Thus, twelve huge concrete locks — larger than any in the World, were soon being built in the jungles of Panama, along with the world's greatest dam (a half-mile wide and 1-1/2 miles long) and a nine-mile cut through a mountain. Such a project had never been undertaken before, anywhere!

The gates of the locks were all mitre-type and all driven by huge electrically-operated gear and pinion mechanisms. Power for the locks was supplied by a hydro-electric generating plant which utilized the 85-foot "head" at Gatun Dam for this purpose and also supplied power to the communities created along the route by the Americans. Electric locomotives were installed to tow all ships through the locks.

The Panama Railroad was rebuilt and relocated so that it would not be submerged by the back-up water from Gatun Dam. Railroad tracks and equipment (developed by Stevens) were an important part of the construction operation, particularly in the Culebra Cut, where huge steam-shovels, and the cars that they loaded — all moved along on tracks as the work progressed. Concrete-mixing and pouring equipment for the locks and dams, of tremendous size, moved on tracks to their work sites. Building materials flowed into the project by rail from huge unloading docks on both the Atlantic and Pacific.

Major troubles were encountered at the Culebra Cut, both during and after construction, when cave-ins and slides developed in the unstable ground of the mountain at that point. But the American workers doggedly cleaned out the huge volumes of extra ground, and widened the cut to keep it open, on each such occasion. During the work at Culebra, a temporary dam was built to prevent the waters of Gatun Lake from spilling over into the cut after the Dam was completed. This was known as 'Gamboa Dyke."

A dramatic celebration took place, on the afternoon of October 10, 1913, when President Woodrow Wilson in the White House, pressed a telegraph key which actuated a huge charge of dynamite, destroying Gamboa Dyke and uniting the waters of the Atlantic and Pacific for the first time. The ceremony took place exactly 400 years to the day (October 10, 1513) after Balboa strode waist-deep into the Pacific Ocean to claim that body of water, and all neighboring countries, for Spain.

The blast touched off by President Wilson, which destroyed the Gamboa Dyke, and let the waters of the Atlantic and Pacific Oceans mingle for the first time, in 1913. ("Panama and the Canal," Willis J. Abbot, 1913)

A large cargo ship being towed, behind the "Electric Mules," between the two locks at Miraflores on the Panama Canal. (Courtesy Phil Cadman.)

The Panama Canal was formally opened to commercial traffic on August 15, 1914. The first ship through the Canal was the U. S. government vessel "Ancon," carrying officials and other guests of honor. In view of the fact that the First World War had begun just days earlier in Europe, the mood was somber and formal celebrations brief. All involved were fully aware of the world-wide significance of the new Canal and the extreme importance of its protection. Fortunately, plans had always included military fortifications at both ends of the Canal, and in recent years, particularly after our entrance into the Atomic age, these plans have been amplified many times over.

The Panama Canal, crowning achievement of any nation at any time in world history, had finally been completed by the United States at a cost of approximately $375,000,000, including our payment to the French company, and the cost of sanitary preparations. The total amount of earth and rock excavated had been 239,000,000 cubic yards. Total length of the Canal — 50.7 miles. Average depth — 45 feet. Height of Gatun Lake, 85 feet above sea level; height of Miraflores Lake, 54-2/3 feet above sea level. Time for a vessel to pass through the Canal — six to eight hours. Distances saved: New York to San Francisco, 7878 miles; New York to Yokohama, 3768 miles; New Orleans to San Francisco 8869 miles; Liverpool to San Francisco, 5666 miles.

A ship entering the Canal from the Caribbean proceeds past Colon, through Limon Bay, a distance of seven miles to the flight of triple locks at Gatun, where it is raised 85 feet, to Gatun Lake. It then travels for 24 miles across Gatun Lake, entering the Culebra Cut (now re-named the Gaillard Cut) at Bas Obispo, traveling nine miles through this artificial channel (300 feet wide at the bottom) to Pedro Miguel Lock. Here the ship is lowered 30-1/3 feet to Miraflores Lake, along which it travels for one and one half miles to the double-flight of Locks at Miraflores. Here it is dropped the remaining 54-2/3 feet to tide-water and proceeds for another 8-1/2 miles, past the city of Panama, to the Pacific Ocean.

In 1938 Madden Dam was constructed further upstream on the Chagres River and about nine miles from the Canal, to provide additional water reserve for the Canal and additional hydro-electric power for the Canal Zone. In 1939 Congress granted funds (in view of Hitler's activities in Europe) for the strengthening of the defenses of the Canal and the construction of a third set of single-chamber canal locks 3000 feet from the original pairs of locks. These plans were never carried out, due to navigational hazards and the sharp turns which would have been required for vessels traveling the new locks. In 1942 a highway was opened between Colon and the city of Panama, to supplement the old Panama Railroad route. Shortly afterward the Navy built twin fuel pipe lines between Cristobal and Balboa.

Plans are still being discussed for the building of a new sea-level canal across Panama, as well as a second canal across Nicaragua.

NAVIGABLE RIVERS

In the early days of inland transportation in America, attempts were made by various of the colonies, later States, to improve river navigation. It was not until 1824, when the Army Corps of Engineers was given the authority to control inland navigation, that significant improvements were possible. In this chapter we would like to review several of the more important river channel improvements in the heartland of the country which have added greatly to its commercial growth in recent years.

THE OHIO

For the past two centuries, the Ohio River has been a major avenue of travel into the West. Since much of the migration was originally downstream, canoes, rafts, keelboats and later — steam boats, floated down the Ohio for many months of the year. Upstream travel was always more difficult. Unfortunately, the level of the river water was far from stable. After long, dry periods the Ohio often became low enough to walk across; boats settled to the mud of the river bottom until the next heavy rain provided enough water for them to continue their journey.

William Milnor Roberts, who had had considerable experience in designing dams and locks for the highly successful Monongahela Navigation Company, not to mention the Welland Canal and various canals in Pennsylvania and Ohio, was employed by the Corps of Engineers in 1866 to study the Ohio and its problems. Most of the Corps' work prior to that time had been clearing of the route of wrecks and snags, dredging the channel, and the building of a short canal to by-pass the falls at Louisville, Kentucky. The shallow, meandering channels had been navigable for the packets and small steam

A lock and dam on the Monongahela River near Fairmount, West Virginia, Circa 1960. (Photo by the Author)

boats, but Roberts pointed out that the deep-water coal barges which were beginning to make their appearance on the River must have wide, relatively straight channels. He recommended the complete "canalization" of the Ohio by a series of dams and locks for the entire length.

Roberts' report was submitted in 1870 to a board headed by Colonel William E. Merrill of the Corps; along with several other reports for bringing water from Lake Erie, creating local reservoirs, and building a completely separate canal, parallel to the river. Merrill favored Roberts' plan, because it had already been successfully applied to the Monongahela, and was also the least expensive.

In 1874, Colonel Merrill recommended to Congress the construction of thirteen locks and movable dams between Pittsburgh and Wheeling, and the construction of Lock Number One at Davis Island, about 5 miles downstream from Pittsburgh. Congress responded with an appropriation of $100,000. After a study of the movable-wicket dam, developed by Jacques Chanoise in France in 1852, Merrill began building a Chanoise-type dam at Davis Island.

This dam was composed of "wickets" (wooden frames 3 feet 8 inches wide and 13 feet long) which collapsed against the dam foundation at high water allowing the barges to pass over it, without locking. When the river level fell, the wickets were pulled into an upright position again, creating slack-water navigation upstream and using the lock to pass the dam. Merritt planned his original lock to be 630 feet long and 78 feet wide, but got violent opposition from the Rivermen, who were running barge tows of four barges abreast (about 100 feet wide). This meant that they would have to break up their tows at each of the more than fifty locks proposed for the Ohio.

Merrill went back to the drawing board and came up with a lock 110 feet wide, by 600 long, which became the accepted "standard" on the Ohio and Mississippi Rivers. However, Merrill's problem with a lock this wide was the construction of mitre-gates ("V-gates") for this tremendous width, which would not collapse under their own weight. He solved the problem by designing a rolling lock gate which ran out of the lock wall, to provide a vertical closure for each end of the lock. Due to the gain in effective lockage length he reduced the lock length from 630 to 600 feet and was able to handle ten coal barges and a towboat without breaking up the tow for double lockage. Merrill's movable dam and track-mounted lock gates proved so successful at Davis Island that they were used for most of the subsequent construction work on the upper Ohio.

The Montgomery Island Lock and Dam on the Ohio River, across river from Beaver, Pa. (Photo by the Author)

In 1910 Congress approved the complete canalization of the Ohio River and by 1929 a system of forty-six locks and dams had been completed — thus assuring year-round navigation of the 981 miles between Pittsburgh and Cairo, Illinois. The navigation was supplemented by reservoirs in the Ohio drainage basin which could be turned into the river during low-water periods.

After World War II, the coming of the more powerful diesel towboats, to replace the old steam towboats, made it possible to push tows twice as long as the previous ones, and — again — improvements were needed to the locking arrangements. In 1955 a modernization program was begun on the Ohio to replace the original forty-six locks with nineteen, higher, gated dams, each dam with a dual lock chamber and at least one 1200-foot long by 110-foot wide lock. As of 1976, eight of the new dam-lock combinations were in operation and the others well along in construction. The total number of locks had been reduced to twenty-six. Even though far from complete, the improvements have brought about a remarkable increase in Ohio River traffic. The 22 million tons carried on the river in 1929 had grown to 136 million tons in 1974.

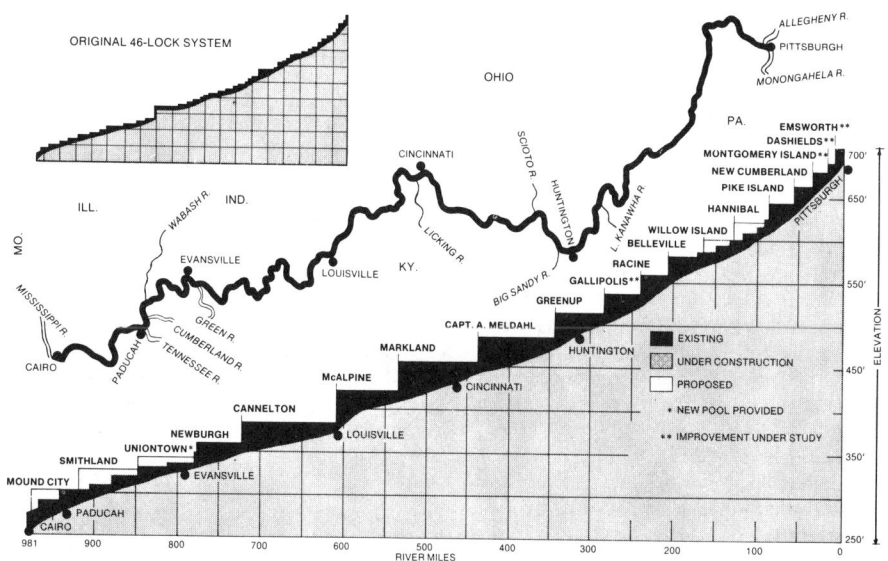

Profile showing the nearly completed replacement of the original 46 locks on the Ohio River with (ultimately) nineteen much larger (110 by 1200-foot) locks, bigger than those on the Panama Canal. The new locks have already reduced lockage time by fifty hours. (Courtesy Engineering News Record)

An Arkansas "stern-wheeler" now operating on the Muskingum River, a tributary of the Ohio, tied up at Zanesville, Ohio. (Photo by the Author)

THE MISSISSIPPI

The Mississippi River, second longest river in the United States, draining approximately two-fifths of the total U.S. land area, has been important historically and commercially since the days of the Spanish and French occupation. It has always been the main line of travel between the Ohio River, the Central States and the great port of New Orleans. With the opening of the Illinois and Michigan Canal, it also began carrying water traffic to and from the Great Lakes.

The Spanish had opened a short canal east from New Orleans in 1785 to avoid the silt-filled delta of the Mississippi River, known as the Carondelet Canal. After the purchase of the Louisiana Territory by the United States in 1803, numerous unsuccessful attempts were made to clear a channel from the Gulf of Mexico into New Orleans sufficiently deep to pass large ocean-going vessels into that port. The first satisfactory channel, thirty feet deep, was engineered by James B. Eads, best known for his bridge across the Mississippi at St. Louis (1873 to 1875). He accomplished his objective by building jetties through the Delta in such a way as to scour the channel with the outflow of the river itself.

After the Civil War, the populace of lower Mississippi was in desperate straits. Ravaged by War the area had never recovered from heavy pre-war floods. The river itself was still choked with the wreckage of both Union and Confederate gunboats and other river-born vessels, sunk during the heavy fighting below and above Vicksburg. River ports had been burned or ravaged by violent engagements on land and water. The area was still heavily-flood prone. In 1874 Congress authorized surveys of the river for both navigation and flood control and in 1879 formed the Mississippi River Commission. The Commission had seven presidential appointees, three from the Army Engineer Corps, with instructions to "deepen the channel, protect the banks, prevent destructive floods, and promote and facilitate commerce." The Commission made its first recommendations in 1880, which called for a complete system of levee and channel improvements, and the work began. After a flood in 1912 levees were raised another three feet, but a severe flood in 1927 proved this was not enough. The 1928 Flood Control Act authorized extensive further improvements to the River, both for flood control and improved navigation channels, between Cairo, Illinois and New Orleans. An experimental station and hydraulic laboratory, using small-scale models of the Mississippi, was built at Vicksburg and Clinton (Mississippi) which has since become the most outstanding installation of its kind in the World.

Meanwhile, between 1884 and 1895, the Army Corps of Engineers constructed five huge dams and reservoirs at the headwaters of the Mississippi to aid in navigation in time of low water, enlarging them and adding a sixth dam-reservoir in 1911. Following a careful study of upper Mississippi improvements in 1930, Congress approved construction of a nine-foot navigation system between Minneapolis and the mouth of the Illinois River. This system included construction of twenty-eight dams and locks (600 by 110-feet) to create full "canalization" of the upper river.

This project has been in operation since 1940, with resulting rapid increase to river traffic on the Mississippi above St. Louis. In 1939 only 2.4 million tons of commodities were barged between Minneapolis and St. Louis. In 1970 this had increased to 54 million tons. Most of the goods transported are petroleum products, coal and grain. Many new harbors, terminals and riverside industries have developed along this portion of the Mississippi. Frequent "bottlenecks" at Lock and Dam Number 26 (Afton, Illinois) have indicated the need for larger locking facilities at that point. River traffic between Cairo

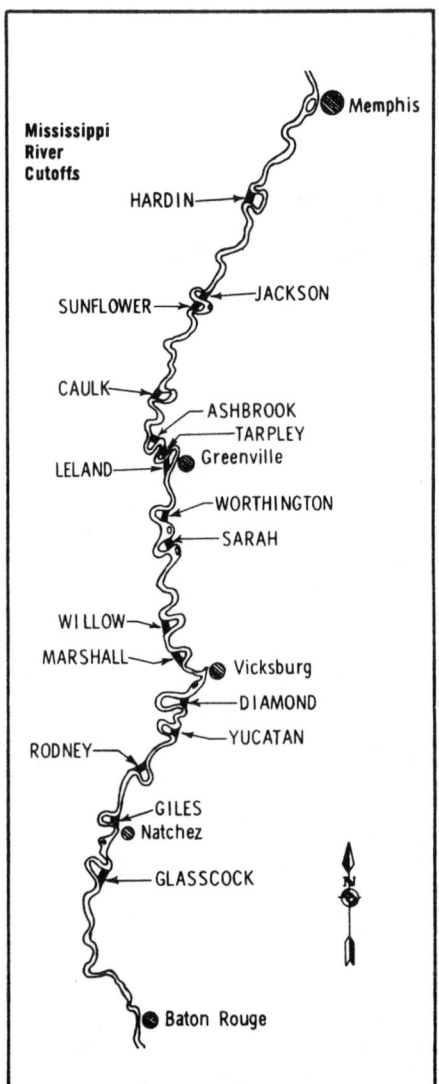

Man-made "cut-offs" on the lower Mississippi. Begun in the 1930's, these sixteen "cut-offs" have reduced total river length by 170 miles. (Courtesy Falk Corporation)

and New Orleans continues to grow as always. A longer navigation season above St. Louis is presently obtained by release of water from the upstream reservoirs to break the ice jams. Much work remains to make the entire Mississippi navigational year-round.

THE TENNESSEE

The Tennessee Valley Authority Act of 1933 provided for the building of a nine-foot navigational channel on the Tennessee River from Paducah, Kentucky to Knoxville, Tennessee, a distance of 650 miles. Two dams already were built; Wilson Dam at Florence, Alabama and Hales Bar Dam near Chattanooga, Tennessee. Seven more high dam-lock combinations were constructed, 1933 to 1944, and a larger main lock (110 by 600 feet) was added at Wilson Dam in 1959. A new dam and lock was built in 1963 on the Clinch River, a subsidiary of the Tennessee, running up toward Oak Ridge, extending navigation to 750 miles.

Since that time enlargements have been made to some of the older locks and a by-pass to the near-by Cumberland River navigation constructed, via Kentucky Lake, upstream from Paducah. In addition, the Tennessee-Tombigbee Waterway, which has been building since 1972, now connects the Tennessee River with the Black Warrior River in Alabama, giving eastern Tennessee a direct water route to the port of Mobile, Alabama and the Gulf of Mexico.

The Tennessee Waterway — open year-round, as well as the inexpensive power generated by the hydro-electric plants at each of the dams, has done a great deal to improve the economy of the entire region through which it passes. TVA acts as the "big brother" of potential new industries and commercial enterprises in the Tennessee Valley and provides moral and even financial support where necessary. The oil companies of Texas have built huge distributing terminals at many points along the river, and are now shipping barge-loads of oil from Houston and Port Arthur into the area. Grain companies, recognizing the expanding livestock industries along the route, have built mills and are shipping corn, soybeans, alfalfa pellets and other agricultural commodities. Commerce in coal, wood and steel is also developing. More than fifty companies operate barges on the Tennessee River. In 1973, 39 million tons of commercial freight moved on the river, setting a new record for the twelfth consecutive year. The same year, industry along the river reported expenditures of $329 million for new or expanded waterfront plants and terminals.

Inside the pilot's cabin of a towboat maneuvering an oil-tanker into one of the locks on the upper Mississippi. (Courtesy "The Texaco Star")

20 GREAT NAVIGATION SYSTEMS

Alabama-Coosa Rivers	305	James River	89
Allegheny River	72	Kanawha River	91
Apalachicola, Chattahoochee, and Flint River System	297	Kentucky River	259
Atlantic Intracoastal Waterway	1,129	McClellan-Kerr Arkansas River Navigation System	448
Black Warrior, Warrior, and Tombigbee River System	466	Mississippi River	2,360
Columbia-Snake Rivers	670	Missouri River	732
Cumberland River	317	Monongahela River	129
Delaware River	129	New York State Barge Canal	522
Green and Barren Rivers	180	Ohio River	981
Gulf Intracoastal Waterway	1,113	Potomac River	113
Houston Ship Channel	50	Sacramento River	145
Hudson River	155	San Joaquin River	127
Illinois Waterway	354	Tennessee River	652

Table of major navigable rivers and mileages in the USA. ("Sea Power" Magazine)

A towboat taking a fleet of barges up the Mississippi. The term "towboat" is a misnomer, as the diesel-powered control boat actually pushes its "tow" from behind. The term had its origin in the days when the barges were towed from the front. (Courtesy Sea Power magazine)

THE ST. LAWRENCE SEAWAY

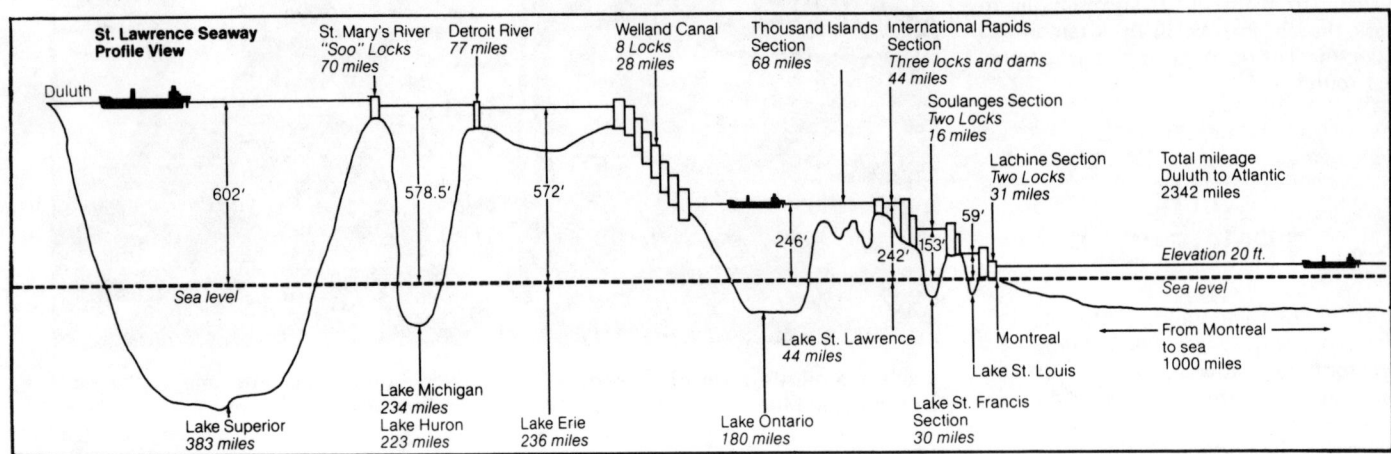

The Canadians have traditionally led the Americans in the development of the present joint-venture on the upper St. Lawrence River, known as the "Saint Lawrence Seaway." Actually the Seaway is only a part of the great inland waterway stretching 2342 miles from the Atlantic Ocean to Duluth on Lake Superior, bringing ocean-going vessels to major lake and river ports of both Canada and the United States.

The first effort of the Canadians to tie this vast navigational system together was the 1797 Sault Ste. Marie Canal joining Lake Superior to Lake Huron. Next came the Lachine Canal (1821-1825) to pass the rapids just above Montreal. Then William Merritt built the first Welland Canal in Canada (1824-1829) making a junction between Lakes Erie and Ontario, around Niagara Falls. The opening of the Cornwall Canal at Long Sault Rapids, the Beauharnois Canal at the Soulanges Rapids, the Williamsburg Canal System and the enlargement of the Welland Canal (1843-50) provided, for the first time, nine-foot navigation all the way from tidewater at Montreal to Lake Erie. By 1887 this entire system had been enlarged again, to provide 14-foot draft navigation.

Headquarters for the American Seaway operating company, as seen from Eisenhower Lock. (Photo by the Author.)

Americans were awakened to the significance of the "Lakes to Sea" route with the discovery of rich deposits of iron and copper ore in the Lake Superior region. In 1855 they built a canal on the American side of the "Soo," to replace the old Canadian Canal which they had destroyed during the War of 1812. Studies were authorized by Congress, in 1841, of commercial shipping possibilities in the entire Great Lakes area.

In 1895 Congress authorized the President to appoint a Deep Waterways Commission to report on the feasibility of building a deep-draft channel from the Great Lakes to the Atlantic Ocean. The

The New Shoreham II, tied up for the night at Prescott, Ontario, is a charter-boat which runs regular trips between Rhode Island, and Canada. In its journey it travels 1200 miles on Long Island Sound, the Hudson River, the Erie Canal, the Oswego Canal, Lake Ontario, the St. Lawrence Seaway, and the Saguenay River. (Photo by the Author.)

Commission met within months with a similar Canadian Commission, recommended a survey for a possible route, and turned the American portion of the job over to the Army Corps of Engineers.

Perhaps if the Panama Canal had not occupied the attention of Congress at the turn of the Century the St. Lawrence Seaway might have been started in the early 1900's. As it was, President William Howard Taft began pushing for a Seaway in 1909, and a succession of American Presidents after him all favored it, but were unable to convince a reluctant Congress — who were under pressure from the railroad interests and east coast port authorities to prevent the project.

In anticipation of the Seaway, the Welland Canal was rebuilt in 1932 — its third major renovation since the original construction — cutting the locks down to eight total (766 x 80 feet) and permitting boats of 30-foot draft to pass through. The bottle-neck was then the upper St. Lawrence, and the pressure for a deep-draft canal at that point grew greater with each passing year.

The entire project was still in limbo during the administration of Franklin Roosevelt, who as governor of New York State clearly recognized the need for the Seaway. As President, he authorized one million dollars (1940) for a further study of the international rapids section of the St. Lawrence by the Corps of Engineers.

After World War II the Canadians were becoming understandably annoyed over what was supposed to have started as a joint venture in 1895. In 1951, Canadian Prime Minister Lester Pearson stated: "The biggest and longest dragging of feet

Lock gates closing behind the New Shoreham II in the Bertrand H. Snell Lock on the St. Lawrence Seaway. (Photo by the Author.)

I have known in my entire career is that of the Americans on the St. Lawrence." Impatient to get the project underway, the Canadian Parliament in December of 1951 passed an Act establishing the St. Lawrence Seaway Authority, for constructing, maintaining and operating — either wholly in Canada or in conjunction with the United States — a deep draft waterway between the Port of Montreal and Lake Erie.

Part of the problem was a decision about the handling of the power which would be generated by the huge dams to be built on both sides of the river. In July of 1953 the New York Power Authority

SEAWAY DATA

There are seven locks in the St. Lawrence River, five in Canada operated by The St. Lawrence Seaway Authority, and two in the United States operated by the Saint Lawrence Seaway Development Corporation. All locks are similar in size. The specifications are:

Length, breast wall to gate fender (Ships may not exceed 730 feet in overall length)	766 feet
Width	80 feet
Depth over sills	30 feet
Locks:	Lift
St. Lambert	15 feet
Cote Ste. Catherine	30 feet
Lower Beauharnois	41 feet
Upper Beauharnois	41 feet
Snell	45 feet
Eisenhower	38 feet
Iroquois	.5 to 6 feet

The locks of the Welland Canal have the same controlling dimensions as those of the Montreal – Lake Ontario Section.

Locks 1-7 of the Welland Canal are lift locks. Lock #8 is essentially a guard lock. Locks 4, 5, 6 are twinned and in flight.

The Welland Canal is 26 miles long and overcomes a difference in level of 326 feet between Lake Ontario and Lake Erie.

The controlling channel dimensions for the Seaway, Lake Erie to Montreal, are:
Depth to a minimum of 27 feet — to permit transit of vessels drawing 26 feet (fresh water draft).
Width of channel:
(a) When flanked by two embankments — 200 feet minimum
(b) When flanked by one embankment — 300 feet minimum
(c) In open reaches — 450 feet minimum

Vessels not exceeding 730 feet overall and 76 feet extreme breadth may transit the Seaway. Vessels' masts must not extend more than 117 feet above water level.

A large freight ship from the Pacific Ocean passes through Lock Number Two on the Welland Canal. (Photo by the Author.)

obtained a license from the Federal Power Commission to construct, operate and maintain the American portion of the power project. This cleared the way for the Canadians to build their portion of the Seaway and cooperate with the New York agency in the hydro-electric development. Finally, in May of 1954, Congress passed the Wiley-Dondero Act which established the St. Lawrence Seaway Development Corporation, authorized to build a 27-foot channel around the international rapids on the American side of the river. Agreements were then reached between the two countries to avoid duplication of navigation facilities. Construction proceeded, with the Army Corps supervising the work on the American side.

The Canadian canals and locks cost $330 million; the canals and locks on the American side, $130 million. The $650 million cost of the hydro-electric facilities was shared equally by the Hydroelectric Power Commission of Ontario and the Power Authority of the State of New York.

As constructed, there are a total of seven locks - 766 feet in length, by 80 in width - to lift vessels 226 feet between mean tide water at Montreal and the level of Lake Ontario. Five of these are on the Canadian side, and two (the Snell and Eisenhower Locks) on the American side. (See separate table.) The greatest lift occurs at Snell Lock, and the least at the Iroquois Lock, which is essentially a guard lock.

Interior of the Eisenhower Lock on the Seaway at Massena, New York. Lift, 38 feet. (Photo by the Author.)

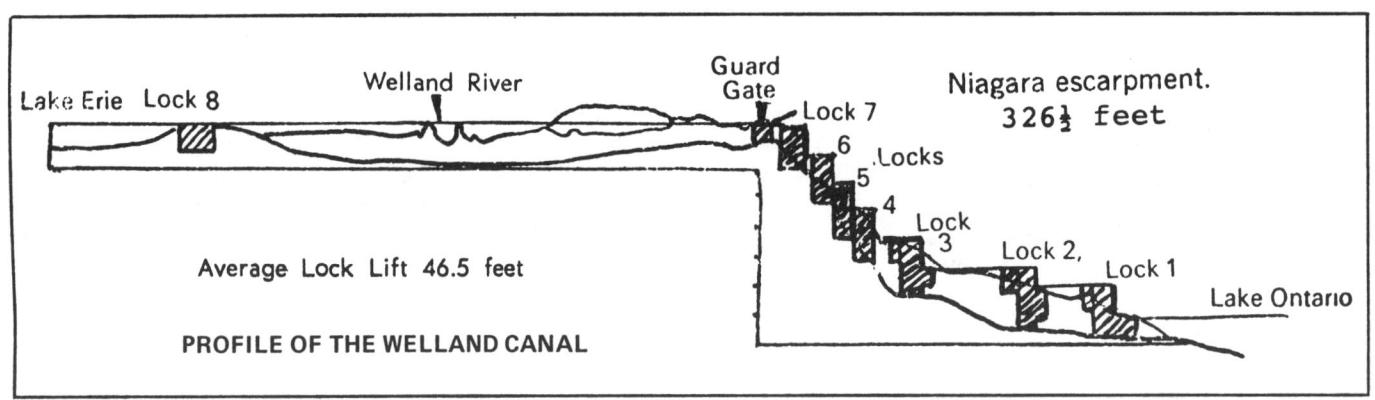

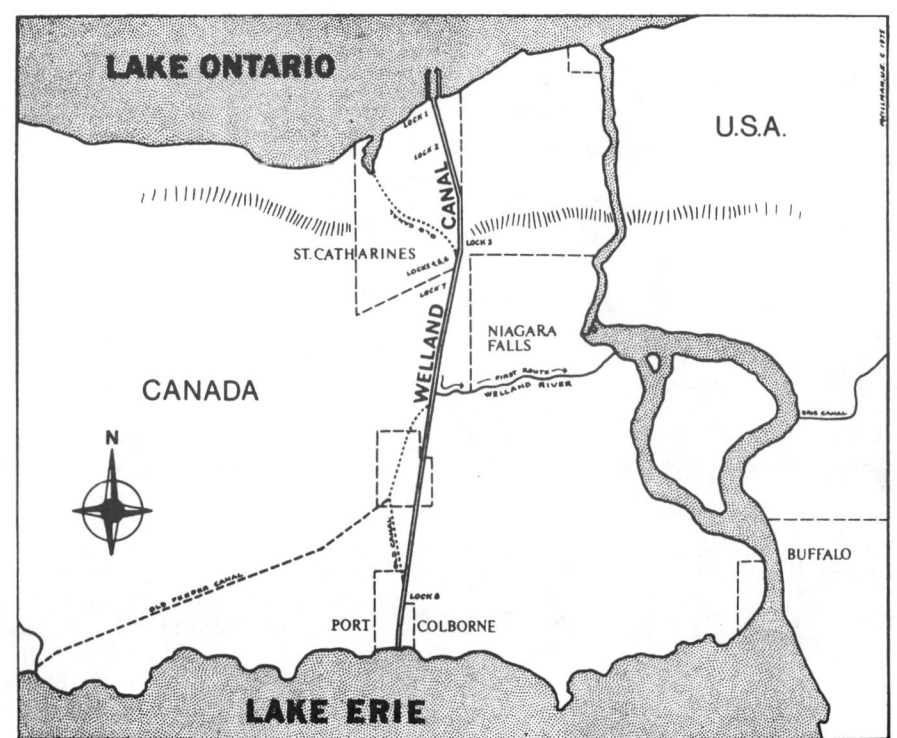

The new Seaway on the St. Lawrence was opened to traffic April 25, 1959. In June 1959, Queen Elizabeth and President Eisenhower formally dedicated the Waterway, which brought to full realization the century-old dream of sailing ocean-going ships into the heart of the American continent!

There are still problems, due to the winter freeze for about four months of the year on the lower St. Lawrence east of Ogdensburg, not to mention the heavy freeze in the Lake Superior district at the other end of the route. However, the entire Seaway, from Duluth to the Atlantic has become one of the great inland waterways of the World, providing a direct connection between the Atlantic Ocean and major cities in the heartland of North America. Ocean-going vessels now dock at such inland ports as Toronto, Buffalo, Erie, Cleveland, Detroit, Chicago, Thunder Bay and Duluth, loading or unloading iron, copper and lead ores; oil, coal, lumber and wheat; and other industrial and agricultural products of two nations.

THE TENN-TOM WATERWAY

The Tennessee-Tombigbee Waterway, first proposed by the French occupants of Southern USA in the 1700's, became a reality in January of 1985! Connecting the navigable upper Tennessee River with the navigable Tombigbee River, it is the most significant development in waterway history in this country since the opening of the Panama Canal and the St. Lawrence Seaway.

To this point in time, all commercial waterway traffic in central United States has been funneled down the Mississippi to the Port of New Orleans, their one and only direct route to the Gulf of Mexico. With the opening of the Tenn-Tom, an alternate route is available for the first time, entering the Gulf of Mexico at the Port of Mobile, Alabama.

The 234-mile waterway, connecting some 16,000 miles of inland waterways on a shortcut from the mid-East region to the Gulf of Mexico, drew years of legal challenges from environmental groups and railroad interests.

Congress scrutinized the project repeatedly during budget hearings, as critics called it a financial boondoggle, with manmade channels ruining scenic waters.

Advocates said it would be an economic bonanza, providing a cheaper barge route for shipments between the Gulf and coal, farm and industrial sites in the lower mid-East, and helping rural Alabama and Mississippi river towns.

It was only in recent years that both Congress and the Federal Courts in Mississippi and Louisiana threw their support behind Tenn-Tom, and its opponents withdrew their suit. With the signing of the final appropriations bill by President Ronald Reagan in July of 1983, the completion of the Tenn-Tom was assured.

Not only does the Tenn-Tom provide water-mileage savings of up to 875 miles by large vessels and "tows" in the Tennessee Valley Authority basin, but it also provides relatively quiescent navigation, with its 110-foot wide locks and slack-water river navigation, as opposed to the sometimes swift currents in the lower Mississippi. This means power-savings for the users of the Tenn-Tom.

TENNESSEE-TOMBIGBEE WATERWAY NAVIGATION CHARTS

Copies of the official navigation charts of the Tennessee-Tombigbee Waterway may be obtained from:

The District Engineer
U.S. Army Engineer District, Mobile
Corps of Engineers
P.O. Box 2288
Mobile, AL 36628

Payment ($14.95) may be made by check or money order payable to: FAO, Corps of Engineers, Mobile.
PAYMENT SHOULD ACCOMPANY REQUEST.

TENN-TOM DESCRIPTION

Length	— 234 Miles
Depth	— Minimum 9 Feet
Width	— 300 Feet
Locks	— 110 Feet x 600 Feet, (10 Locks)
Barges	— 8 barge tow configuration. Speed: 6.0 mph average. Tow size: 585 feet x 105 feet x 8½ feet draft.

Scene at the Columbus, Mississippi Lock and Dam on the Tenn-Tom, January 18, 1985. The "Tow" of the Waxler Towing Company of Memphis is shown edging its way into the 100-foot wide by 600-foot long lock, which is standard for all ten locks on the Tenn-Tom.

TENNESSEE-TOMBIGBEE WATERWAY TYPICAL MILEAGE SAVINGS

MILEAGE CHART

Ports VIA:	Mississippi River (miles)	Tenn-Tom (miles)	Distance Savings (miles)
Pensacola to Chattanooga	1541	771	720
Mobile to Paducah	1061	666	395
Pittsburgh to Birmingham	2396	1567	829
Gulfport, MS to Decatur, AL	1229	628	601
Chicago to Mobile	1567	1256	311
Nashville to Mobile	1266	871	395

Construction of the Tenn-Tom waterway was begun by the Corps of Engineers in December 1972 and completed December 1984, 21 months ahead of schedule.

Construction costs of the Tenn-Tom were on the order of $1.9 billion, in State and Federal funding. Its construction was divided into three sections: a 149-mile river navigation section in Alabama in the south; a 46-mile "canal" section further north; and (the most difficult) a 39-mile "divide" section through the Tennessee Divide, a rise that separates the Tennessee Valley from the Tombigbee Valley.

The northern end of the Tenn-Tom is on the Pickwick Lake portion of the Tennessee River and near the common borders of Alabama, Mississippi and Tennessee. At its southern end the waterway connects at Demopolis, Ala, with the Black Warrior River which flows southward to the Gulf of Mexico at Mobile.

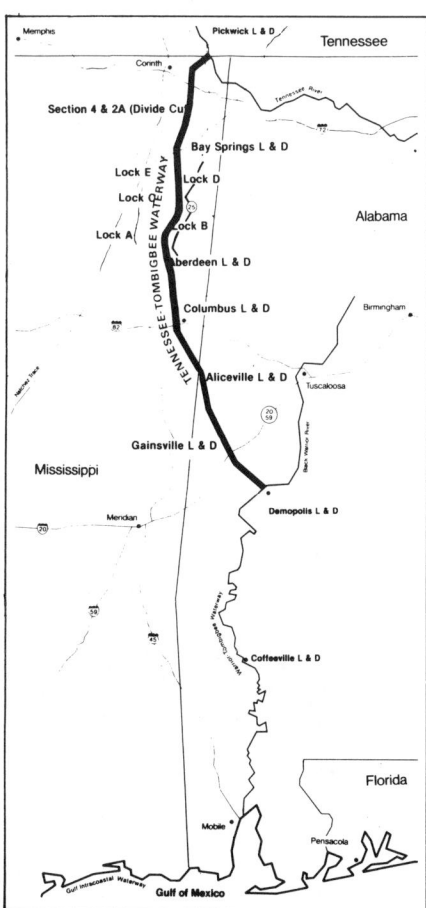

The official dedication of the new Waterway took place June 1, 1985 at the Columbus Lock and Dam, Columbus, Mississippi.

HISTORIC CANAL PRESERVATION

In recent years there has been rapidly growing interest in the preservation of the remaining ruins and relics of the great 4500-mile canal system of the 1800's in northeastern United States.

At this writing, a total of forty-three non-profit historical organizations in Northeast United States, and six in Canada, are assisting in the selection of historic canal sites to be placed on the National Register; publishing information about historic canals in news-letter and booklet form; conducting field trips along old canal routes; holding seminars; and re-watering and re-opening short sections of historic canals as park areas. Almost every one of the northeastern States, where canals carried passengers and freight 100 years ago, now has its own "Canal Society."

Sectional Canal Freight Boats, used on the Allegheny Portage and Philadelphia and Columbia Railroads, are illustrated in this model, in the Lemon House Museum at Cresson, PA.

The "St. Helena II," first canal-boat replica built in Ohio, wends its leisurely way along a section of the Ohio and Erie Canal near Canal Fulton. (Courtesy Roadway Express Magazine.)

The Delaware and Raritan Canal is one of the few historic canals that has never gone dry since its construction in 1834. The State now maintains it as a source of water for industries and small communities in lower New Jersey. (Photo by the Author.)

Restored Lock and re-watered canal channel on the Chesapeake and Ohio Canal at Glen Echo, Maryland. (Photo by the Author.)

Packet-boat replica "General Harrison" loading for a trip along the Miami and Erie Canal in the Piqua Historical Area. (Courtesy Ohio Historical Society.)

Canal Packet Boat Model at Lemon House, Cresson, Pa. This Museum is operated by the National Park Service, and includes many displays depicting operations of the Allegheny Portage Railroad.

Replica of a packet boat — the "Colonel Baldwin" — is towed by horses along a re-watered section of the old Middlesex Canal in Woburn, Massachusetts. (Courtesy Historical Commission of Woburn.)

The American Canal Society, with headquarters in York, Pennsylvania, has currently been providing a regular medium for the exchange of canal information on an inter-national basis; and a Canadian Canal Society has been formed in St. Catharines, Ontario.

At a number of points in New England, New York State, Pennsylvania, Maryland, Ohio and Indiana it is now possible to ride a canal boat replica, back of mules or horses, along short sections of restored canals, just as you would have done it 150 years ago. Major collections of canal utensils, boat furniture, canal boat models, photographs and other memorabilia have been carefully gathered into various canal museums in the Northeast.

On these pages some of these museums and restored canal sections are shown in photos, taken by the author and other canal buffs, who are unwilling to let our historic canal heritage become a musty record on the shelves of Time and are endeavoring to bring it all back to life.

The "Monticello II" canal-boat replica prepares to tow a load of passengers along a re-watered section of the Ohio and Erie Canal at Roscoe Village, an old canal town in Ohio. (Photo by the Author.)

The Canal Museum at Easton, Pennsylvania, operated jointly by the Hugh Moore Park and the Pennsylvania Canal Society. Located at the junction of the Lehigh and Delaware Canals. (Photo by the Author.)

The "Ben Franklin," an authentic canal boat replica, is shown in a restored lock at the south end of Metamora, Indiana. Nearly ten miles of the Whitewater Canal have been re-watered on both sides of this old canal town. (Photo by the Author.)

Working model of Inclined Plane Number Six on the Portage Railroad at Lemon House Museum. Engine house to the left; Lemon House at the center. (Photo by the Author.)

Lock Number 50 on the eastern section of the Sandy and Beaver Canal, near the canal town of Frederickstown, Ohio, has been dug out of the mud and completely restored, with mitre-gates which work. (Photo by the Author.)

The "Josiah White," a freight-boat replica, carries a load of passengers along a three-mile re-watered section of the Lehigh Canal above Easton, Pa., back of a mule team. (Courtesy Hugh Moore Park.)

This is what passengers see from the bow of a canal boat on the Delaware Canal at New Hope, Pa. In the foreground is a short aqueduct which they are about to cross. Note the tow-line and mule team out in front. The sixty-mile Delaware Canal has run "bank-full" of water for the past 150 years! (Photo by the Author.)

Freight Boat Number 249, of the Lehigh Canal and Navigation Company, was recently raised by canal buffs from the bottom of a water-filled quarry in Northampton County, Pa. It had been sunk there sixty-five years ago when a near-by section of the Lehigh Canal ceased operation. (Photo by Dan Urich for the Allentown Morning Call.)

FUTURE INLAND WATER TRAVEL

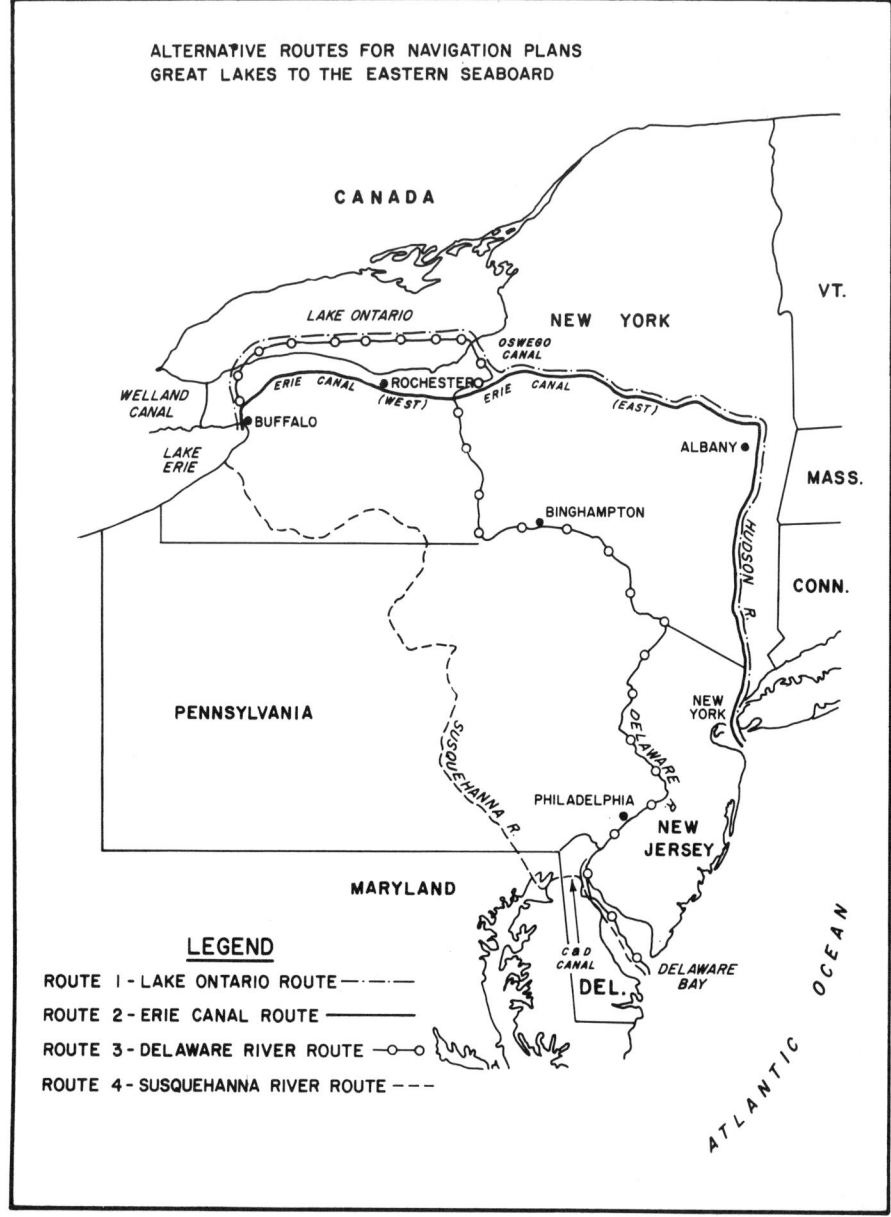

The Corps of Engineers is investigating routes for a deep-draft ship canal from the Great Lakes to the Eastern Seaboard, using an All-American, All-Weather route. Four possibilities are shown here. (Army Corps map.)

With the present energy crisis sending us in search of new and less expensive means of transportation, our inland waterways provide great hopes for the future. Water transport has always proven far more economical than railroad, truck or air transport for the movement of heavy goods and products where speedy delivery is not a problem. The Corps of Engineers continue to investigate new major water routes to supplement our present ones. For instance, they are currently studying an "All-American, All-Weather Route" from Lake Erie to the Atlantic, which may cut through the heart of Pennsylvania. (See sketch). Also under consideration is a new Isthmian Canal, possibly at sea level, which may cut across another section of Panama, or use the old route in Nicaragua. By the year 2000 we should see radical changes in our entire transportation pattern, which will include many new uses and new additions to our coastal and inland waterways system.

A non-profit organization known as the "Great Lakes Inland Waterway Authority", with headquarters in Youngstown, Ohio proposes a ship-canal between Beaver Falls, Pa and Painesville, Ohio. The accompanying map shows how the proposed canal would create a "complete water circle" in central USA to provide inexpensive water transit throughout this heavily industrialized area.

BIBLIOGRAPHY

Abbott,, Willis J. - "Panama and the Canal" (1913)

American Canal Society - "Best from American Canals" (1984)

American Public Works Association - "History of Public Works in the United States" (1976)

American Society of Civil Engineers - "Biographical Dictionary of American Civil Engineers (1972)

American Society of Civil Engineers - "The Civil Engineer, His Origins" (1970)

American Society of Civil Engineers - "Civil Engineering History" (1978)

Andrist, Ralph K. - "The Erie Canal" (1964)

Anness, Milford - "Low Bridge and Locks Ahead" Whitewater Canal (1972)

Baer, Christopher - "Canals and Railroads of the Mid-Atlantic States 1800-1860" (1981)

Baker, Ira O. - "A Treatise on Masonry Construction" (1906)

Basset, John M. & Petrie A. R. - "William Hamilton Merritt" (1974)

Bishop, James B. "The Panama Gateway" (1913)

Boucher, Cyril - "James Brindley" (1972)

Brown, Alexander Crosby - "The Dismal Swamp Canal"

Bracegirdle, Brian and Miles, Patricia H. - "Thomas Telford" (1973)

Cain, Louis P. - "Sanitary District of Greater Chicago" (1980)

Clarke, Mary Stetson - "The Old Middlesex Canal" (1974)

Condon, George - "Stars in the Water" (1974)

Conservation Dept., State of Illinois - "Illinois & Michigan Canal State Trailway" (1974)

Cummings, Hubertis M. - "Pennsylvania Board of Canals Commission Records" (1959)

Dunaway, Weyland Fuller - "History of the James River and Kanawha Co."

Dunbar, Seymour - "History of Travel in America" (1937)

Fisher, Allan C., Jr. - "American Inland Waterways" (1973)
Fatout, Paul - "Indiana Canals" (1972)
FitzSimons, Neal - "Reminiscences of John B. Jervis" (1971)
Flexner, James Thomas - "Steamboats Come True" (1978)
Gallatin, Albert - "Public Roads and Canals" - 1808 (1968 reprint)
Gard, R. Max - "The Sandy and Beaver Canal" (1952)
Godfrey, Frank H. - "The Godfrey Letters" (1973)
Gray, Ralph - "The National Waterway" (1967)
Hadfield, Charles - "The Canal Age" (1968)
Hadfield, Charles - "Afloat in America" (1979)
Hahn, Thomas F. - "George Washington's Canal at Great Falls, Va." (1976)
Hare, Jay V. - "History of the Reading (R.R.)" (1966)
Harland Bartholemew Assoc. - "Illinois & Michigan Canal Development Plan" (1974)
Harlow, Alvin F. - "Old Towpaths" (1926)
Hepburn, A. Barton - "Artificial Waterways of the World" (1909)
Heisler, John P. - "Canals of Canada" (1973)
Historical American Buildings Survey - "Georgetown Architecture - The Waterfront" (1968)
Holton, Gladys Reid - "The Genessee Valley Canal" (1971)
Howe, Walter A. - "Documentary History of the Illinois and Michigan Canal" (1956)
Hyde, Charles K. - "Upper Pennisula of Michigan" (1978)
Jackson, John N. - "Welland and the Welland Canal" (1975)
Jacobs, David L. & Neville, Anthony E. - "Bridges, Canals & Tunnels" (1968)
Judson, Clara Ingram - "St. Lawrence Seaway" (1959)
Kemp, Emory L. - "SIA Journal, Volume Two, Number One" (1976)
Kirkwood, James I. - "Waterway to the West" (1963)
Kulik, Gary - "Rhode Island Inventory of Historic Engineering and Industrial Sites" (1978)
Langbein, W. B. - "Hydrology and Environmental Aspects of the Erie Canal" (1976)
Lee, James - "The Morris Canal" (1974)
Legget, Robert - "The Canals of Canada"
Legget, Robert - "Rideau Waterway" (1972)
LeRoy - "The Delaware and Hudson Canal" (1980)
Lewis, Gene D. - "Charles Ellet, Jr. 1810 - 1862" (1968)
Lewis, M. J. T. - Slatcher, W. N. & Jarvis, P. N. - "Flashlocks on English Waterways" (1969)
Livingood, James W. - "Philadelphia-Baltimore Trade Rivalry" 1780-1860 (1947)
Lossing, Benson J. - " A History of the United States" (1867)
Ludwig, Edward J. III - "The Chesapeake and Delaware Canal" (1979)

Marshall, Logan - "Story of the Panama Canal" (1913)
Mayhill, Dora Thomas - "Old Wabash and Erie Canal in Carroll County" (1953)
Maryland Historical Society - "Virginia Journals of Benjamin Henry Latrobe, 1795-1798" (1977)
McCulloch, David - "The Path Between the Seas" (1977)
McCullough, Robert & Leuba, Walter - "The Pennsylvania Main Line Canal" (1976)
McKelvey, William J., Jr. - "Champlain to Chesapeake" (1978)
McKelvey, William J., Jr. - "The Delaware & Raritan Canal" (1975)
Miller, John P. - "The Lehigh Canal" (1979)
Mitchell & Hinman - "Internal Improvements in the USA - 1835" (1972)
Morton, Eleanor - "Josiah White, Prince of Pioneers" (1946)
Myer, Donald B. - "Building the Potomac Aqueduct" (1975)
O'Donnell, Thomas C. - "Snubbing Posts" (1972)
Patterson, Wallace and Patrick, Sam - "Presidents of Manifest Destiny" (1973)
Payne, Robert - "The Canal Builders" (1959)
Pennsylvania Canal Society - "History of the Monongahela Navigation 1873" (1978 reprint)
Phillips, John - "Inland Navigation, 1792" (1970)
Porcher, F.A. - "The Santee Canal" (1970)
Potterf, Rex M. - "Wabash and Erie Canal" (1970)
Roberts, Christopher - "The Middlesex Canal" (1938)
Ryan, David D. - "Falls of the James" (1975)
Sackheim, Donald E. - "Historical American Engineering Record Catalog for 1976"

Sanderlin, Walter S. - "The Great National Project: A History of the Chesapeake and Ohio Canal" (1946)
Sanderson, Dorothy H. - "Delaware and Hudson Canalway" (1972)
Scheiber, Harry N. - "Ohio Canal Era" (1960)
Shank, William H. - "The Amazing Pennsylvania Canals" (1981)
Shank, William H - "300 Years with the Pennsylvania Traveler" (1976)
Smeltzer, Gerald - "Canals Along the Lower Susquehanna" (1962)
Smith, Peter L. - "Canal Barges and Narrow Boats" (1975)
Snyder, Frank E. & Guss, Byron H. - "The District (C.E.)" (1974)
Squires, Roger W. - "Canals Revived" (1979)
Swanson, Leslie C. - "Canals of Mid-America" (1964)
Swetnam, George - "Pennsylvania Transportation" (1968)
Tanner, Henry S. - "Canals and Railroads of USA 1840" (1970 reprint)
Taylor, George Rogers - "The Transportation Revolution 1815-1860" (1951)
Trevorrow, Frank - "Ohio's Canals" (1973)
U. S. Army Engineer Institute for Water Resources - "National Waterways Study - (Maps)"
Veit, Richard F. - "The Old Canals of New Jersey" (1963)
Wakefield, Manville B. - "Coat Boats to Tidewater" (1965)
Werner, Constance W. - "Georgetown Historic Waterfront" (1968)
Western Writers of America - "Water Trails West" (1978)
White, Josiah - "Josiah White's History by Himself" (undated)
Woods, Terry K. - "Twenty Five Miles to Nowhere" (1978)
Yoder, C. P. - "Delaware Canal Journal" (1972)

A mixed group of travelers appear to have commandeered this canal freight boat for a short trip down the Eastern Division of the Main Line into Harrisburg, Pennsylvania. The boat is shown emerging from the outlet lock on Duncan's Island, just before crossing to the east bank of the Susquehanna. Note the mule team on the towing path of the old Clarks Ferry Bridge in the background. (Courtesy James A. O'Boyle.)

INDEX

Allegheny Portage Railroad 66,67,68
Alexandria Canal 26
Albemarle & Chesapeake Canal 30
Allen, Horatio 33
American Canal Society 1,67
Army Corps of Engineers 49,69

Balboa, Vasco Nunes de 51
Baldwin, Loammi 16,17
Baltimore & Ohio RR 26
Barton Aqueduct 7
Bates, David Stanhope 31
Beauharnois Canal 43,62
"Ben Franklin" Canal Boat 68
Bibliography 69,70
"Big Chute" 43
Birmingham & Liverpool Canal 10
Bridgewater Canal 6
Brindley, James 6,7,8

Caer Dyke 6
Caledonia Canal 10
Canadian Canals 41,42,43,44
Canadian Canal Society 67
Canal du Midi 3
Canal Engineering 37
Canal Engineers 36
Canals of Antiquity 3
Carillon Canal 42
Carondelet Canal 15,60
Chambly Canal 19,43
Champlain Canal 19
Chanoise, Jacques 59
Chesapeake & Delaware Canal 27
Chesapeake & Ohio Canal 26,27,66
Chicago Sanitary & Ship Canal 46
Chinese Canals 4
Chockoyette Aqueduct 34
Clayton-Bulwer Treaty 51,54
"Clermont" Steamboat 12
Clinton, DeWitt 17,18,19
Coal-Carrying Canals 40
"Colonel Baldwin" Canal Boat 67
Columbia 54
Conewego Canal 15
Corinth Canal 4
Cornwall Canal 43,62
Culebra Cut 57
"Cut-offs," Mississippi 60

Damascus Canals 4
Delaware Canal Boat 68

Delaware & Hudson Canal 32,33
Delaware & Raritan Canal 27,28,29,66
DeLesseps, Ferdinand 3
Dismal Swamp Canal 13,14,15
Dutch Canals 5

Eads, James B. 60
Early British Canals 6
Eisenhower, Dwight 64
Elizabeth, Queen of England 64
Ellet, Charles Jr. 25,35,36
Erie Canal 17,18,19
Foss Dyke 6
Franklin, Benjamin 11
French Canal, Europe 5
French Canal, Panama 52,53
Fulton, Robert 7,11,12

Gamboa Dyke 57
Gatun Dam 55
Gatun Lake 57
Geddes, James 20
"General Harrison" Canal Boat 67
German Canals 5
Gill, Edward Hall 34,35
Goethals, George W. 56,57
Goeth, Johan Wolfgang 51
Gooding, William 44
Gorgas, William Crawford 55
Göta Canal 8,10
Grand Canal 4
Grand Trunk Canal 7
Greek Canals 4
Grenville Canal 42

Harecastle Tunnel 8
Hay-Pauncefote Treaty 54
Historic Canal Preservation 66,67,68
Hugh Moore Park Museum 67
Hydraulic Canal Cement 39
Illinois & Michigan Canal 44,45,46,47
Illinois Waterway 47
Indiana Canals 30,31
Italian Canals 5

James River Company 14,15
James River & Kanawha Canal . . . 29,34,35,36
Jefferson, Thomas 11,14
Jervis, John Bloomfield 32
"Josiah White" Canal Boat 68
Lachine Canal 41,42,62
Languedoc Canal 3,5
Latrobe, Benjamin Henry 15
Lehigh Canal 28,68

Little Falls Canal 15
Livingston, Chancellor Robert 12
Lockport Flight of Locks 21

"Main Line" Canal 25,33,70
Manchester Ship Canal 7
Maryland Canals 26
Massachusetts Canals 23
Merrill, William E. 59
Merritt, William Hamilton 42,62
Miami & Erie Canal 24
Middlesex Canal 16
Mississippi Navigation 60,61
"Monticello II" Canal Boat 67
Morris Canal 27,28
Navigable Rivers 58,59,60,61
New Jersey Canals 27
Nicaraguan Canal 51,54
Ohio Canals 23,24
Ohio & Erie Canal 23
Ohio Navigation System 58,59

Panama Canal 57,58,69
Panama Canal Company 52
Panama Canal Profile 56
Panamanian Revolution 54
Patowmack Co. 14,15
Pennsylvania Canals 24,25
Penn, William 11
Peterborough Lift Lock 44
Pien Canal 4
Poe, Orlando M. 50
Pontcysyllte Aqueduct 9
Portage Canal 48
Pound Lock 4

Red Flag Canal, China 4
Rideau Canal 42
Roanoke Canal 34
Roberts, Nathan 20
Roberts, William Milnor . . . 25,33,34,58,59
Roman Canals 4
Roosevelt, Franklin 63
Roosevelt, Theodore 52,53,54

Sandy & Beaver Canal 34,67
Santee & Cooper Canal 15
Sault St. Marie Canal 41,50,62
Schuylkill Navigation 28,35
Senf, John Christian 15,16
"Soo" Canal 41,50,62
South Hadley Canal 15
St. Anne Canal 42
"St. Helena II" Canal Boat 66
St. Lawrence Seaway 62,63,64
St. Lawrence Seaway
 Development Corp. 62,63
St. Mary's Falls Canal 41,50,62
Stanhope, Earl of 12
Strickland, William 24
Suez Canal 3,52
Susquehanna Canal 15
Susquehanna & Schuylkill Canal 14,15
Susquehanna and Tidewater Canal 27

Taft, William Howard 56,63
Telford, Thomas 8,9,10
Tennessee Navigation 61
Tennessee-Tombigbee Waterway 61,65
Trent-Severn Waterway 43
True, John 6
Union Canal 14
Virginia Canals 29,30

Wabash & Erie Canal 30
Wallace, James Findley 55
Washington, George 13,14
Watson, Elkanah 7,12,13,14
Weigh Lock 40
Welland Canals 42,63,64
Weston, William 16,17
White, Canvass 7,22,25
Whitewater Canal 31
Wiley-Dondero Act 1954 63
Williamsburg Canal 43,62
Wilson, Woodrow 57
Wright, Benjamin 19

A watered section of the Union Canal, showing Lock Number 21-west. Directly behind the lock is the ruin of an Aqueduct across the Swatara Creek. (Photo by the Author.)

OTHER PUBLICATIONS OF THE AMERICAN CANAL AND TRANSPORTATION CENTER

THE AMAZING PENNSYLVANIA CANALS — 150th Anniversary Edition — By William H. Shank (1981). A much expanded variation of many previous printings. 125 illustrations, and tables of locks and mileages on most of the principal canals in the State, never previously gathered together in one volume. Four-color cover; two-color interior; 128 pages; a definitive work.

THREE HUNDRED YEARS WITH THE PENNSYLVANIA TRAVELER — By William H. Shank, (1976). A delightful and fascinating book — the entire history of transportation in the Keystone State. No means of travel has been overlooked — Indian trails, early roads and vehicles, river travel, pioneer bridges, canals, aqueducts, inclined planes, gravity railroads, steam railroads, plank roads, bicycles, horse cars, cable cars, trolley cars, early automobiles, pioneer air vehicles, Two-color, 8½ x 11 paperback, the book contains 113 old photos, 89 sketches, 11 maps and two hundred pages.

TOWPATH GUIDE TO THE CHESAPEAKE AND OHIO CANAL — By Thomas F. Hahn, (1984). A fully illustrated, historical commentary and mile-by-mile directory for the entire 184-mile length of the C. & O. Canal Towpath from Washington, D.C. to Cumberland, Maryland. Excellent maps included. 226 pages.

HISTORIC BRIDGES OF PENNSYLVANIA — By William H. Shank, (1980 Edition). Traces the development of the bridge-building arts from the time of the first covered bridge in America, built in Philadelphia in 1805, to modern bridges of the 20th Century. Biographies of such famous bridge builders as John Roebling, Theodore Burr, Charles Ellet and Ralph Modjeski included. Profusely illustrated.

CHESAPEAKE AND OHIO CANAL OLD PICTURE ALBUM — By Thomas F. Hahn (1985). One hundred excellent full-page photographs from the 1800's show peak operations along the C. & O. Canal, from Georgetown to Cumberland, Maryland. The introduction includes a brief history of the canal. Much additional historical information appears in the photo captions.

GREAT FLOODS OF PENNSYLVANIA — A TWO-HUNDRED YEAR HISTORY — W. H. Shank, (Fifth Printing, 1981). Data, photos and non-technical text on all major floods in the Keystone State since records have been kept. A definitive work.

SYLVESTER WELCH'S REPORT ON THE ALLEGHENY PORTAGE RAILROAD, 1833 — Detailed description, by its chief engineer, of one of the most unusual railroads ever built in the United States. A series of ten, steam-powered, inclined planes were part of this amazing railroad. Illustrated with Hoffmann drawings. Fold-out map included, (Fourth printing, 1983.)

THE C & O CANAL — AN ILLUSTRATED HISTORY — By Thomas F. Hahn & Diana Suttenfield-Abshire (1981). A full-size, illustrated sketch book of historic scenes and engineering features along the entire 184-mile length of the C & O Canal. 84 pages of excellent sketches and explanatory captions.

VANDERBILT'S FOLLY — A HISTORY OF THE PENNSYLVANIA TURNPIKE — W. H. Shank, (Seventh Printing, 1982). The railroad war of 1880-85 which created the tunnels and roadbed for the present turnpike. History of the Turnpike, 1940-1982, included.

INDIAN TRAILS TO SUPERHIGHWAYS — By William H. Shank, (1982 Printing.) History of the development of Pennsylvania's historic roads and the many interesting vehicles used on them. Much Indian folklore and early colonial history. Descriptions of Braddock's Road, Forbes' Road, National Highway, Lancaster Turnpike, Plank Roads, Corduroy Roads, William Penn Highway, Lincoln Highway, Pennsylvania Turnpike and Keystone Shortway. Profusely illustrated.

THE C. & O. CANAL BOATMEN, 1892-1924 — By Thomas F. Hahn (1980). The Life and Times of the men who worked the canal boats in the declining years of the canal's operation. Prepared after careful study of tapes and written interviews with boatmen, many of whom have since passed on. The book abounds with direct quotations from many of them.

JOURNEY THROUGH PENNSYLVANIA — 1835. Edited by William H. Shank (1981). Reprint of a book published by Philip Nicklin of Philadelphia in 1836, describing in detail a trip he made across Pennsylvania by canal, rail and stage coach. His alliterative original title was "A Pleasant Peregrination Through the Prettiest Parts of Pennsylvania Performed by Peregrine Prolix". Illustrated with Hoffmann and Storm sketches, as well as old photographs.

HISTORY OF THE YORK-PULLMAN AUTOMOBILE, 1903-1917 — By William H. Shank, (1970). History of the "Six-Wheeler" Pullman, and its successors, which almost made York, Pa. the automotive capital of the United States. History of the early automotive industry in Eastern Pennsylvania also included. Profusely illustrated. An ideal gift for antique car buffs.

ELLET AND ROEBLING — By Donald Sayenga, (1983). The story of the interplay of the lives of these two famous canal and bridge-builders of the 1800's, and their magnificent suspension structures, some of which still stand today. Fully illustrated.

WHEN HORSES PULLED BOATS — By Alvin F. Harlow. A 1983 reprint of Harlow's little-known canal book, written for school students in 1936. An excellent exposition of the historic canal era in the USA. Introduction by William H. Shank. Canal bibliography included. Well illustrated with sketches by Orson Lowell and Philip Hoffmann.

THE CANALLER'S SONG BOOK — By William Hullfish, Music Instructor, Writer, Singer (1984). This 88 page, 8½ x 11 book contains forty historic canal ballads (words and music) collected from old records in the northeastern USA over a ten year research period. Said to be the most comprehensive collection of canal songs ever published. Illustrated with lively sketches of life on the canals.

THE BEST FROM AMERICAN CANALS — Number I (1972-1979). A publication of the American Canal Society, (third printing 1983) featuring articles written by canal-buff experts on historic and currently operating canals in the U.S.A., Canada and Europe. An 8½ x 11 paperback, with 88 pages, the book contains 150 illustrations and a worldwide index. Nine pages on the Canadian Canals; fifteen pages on the Canals of Europe.

THE BEST FROM AMERICAN CANALS — Number II (1980-1983). A publication of the American Canal Society (1984) reprinting articles on historic and currently operating canals in USA and overseas. Featured are the Panama Canal, the Trent-Severn Waterway, and canals in Europe, Thailand and China. 140 illustrations and maps included, and a world-wide index. An 8½ x 11 paperback with 88 pages.

THE COLUMBIA-PHILADELPHIA RAILROAD AND ITS SUCCESSOR — William Hasell Wilson, 1896. This book is an on-the-spot account of the building of one of the oldest railroads in America by its chief engineer, later resident engineer for the Pennsylvania Railroad, who purchased it from the Pennsylvania Canal Commissioners. This 1985 reprint is fully illustrated with 45 old photos, maps and drawings from the files of William H. Shank.

Inquiries may be directed to the American Canal and Transportation Center, 809 Rathton Road, York, Pa. 17403. Price list and discount schedule available.